VOLUME VI

SIAM–AMS PROCEEDINGS

Stochastic Differential Equations

AMERICAN MATHEMATICAL SOCIETY
PROVIDENCE, RHODE ISLAND
1973

PROCEEDINGS OF A SYMPOSIUM IN APPLIED MATHEMATICS
OF THE AMERICAN MATHEMATICAL SOCIETY AND THE
SOCIETY FOR INDUSTRIAL AND APPLIED MATHEMATICS

HELD IN NEW YORK CITY
MARCH 29–30, 1972

Edited by
JOSEPH B. KELLER
HENRY P. McKEAN

Prepared by the American Mathematical Society under ONR grant NONR(G)-00004-72

Library of Congress Cataloging in Publication Data
Main entry under title:

Stochastic differential equations.

(SIAM-AMS proceedings, v. 6)
Sponsored by the American Mathematical Society and the Society for Industrial and Applied Mathematics and held in New York City, Mar. 29–30, 1972.
"Prepared by the American Mathematical Society under ONR grant NONR (G)–00004–72."
Includes bibliographical references.
1. Stochastic differential equations -- Congresses.
I. Keller, Joseph Bishop, 1923– ed. II. McKean, Henry P., ed. III. American Mathematical Society.
IV. Society for Industrial and Applied Mathematics.
V. Series: Society for Industrial and Applied Mathematics. SIAM–AMS proceedings, v. 6.
QA274.23.S74 519.2′4 72–13266
ISBN 0–8218–1325–0

AMS (MOS) subject classifications (1970). Primary 60HXX

Contents

Foreword

The papers collected here are extremely varied. Our stated task was to bring together a meeting on stochastic differential equations. Usually a pure mathematician will understand this subject to be part of the integral and differential calculus based upon the Brownian motion, dealing chiefly with problems like $\dot{x} = e(x)b^{\cdot} + f(x)$ with known coefficients e and f and a standard "white noise" $b^{\cdot}$. It was, however, our policy to interpret the field much more broadly and to admit any kind of problem in which a randomness enters either via the coefficients or the forcing or both, a simple example being $\ddot{x} + (\varepsilon + \alpha^2)x = \cos \beta t + \gamma b^{\cdot}$, in which ε is a (small) stationary noise and $\alpha\beta\gamma$ are known constants.

The general situation can be modelled as follows: you have the celebrated black box with its input and its output, and all three (box, input, and output) are permitted to involve some element of chance. You can ask

(1) to describe (pathwise or statistically) the output knowing the input, or the other way around;
(2) to describe the box knowing the input and the output (problem of analysis);
(3) to make a box which will convert a specific input into a specific output (problem of synthesis).

Clearly, this is a vast subject with connections to the most diverse fields of application and could be treated in our meeting in the most episodic manner only. Correspondingly, our aim is modest namely, to indicate the extraordinary variety of such problems, especially as they arise from practical considerations, and to hope to influence some of the mathematical audience (both listeners and readers) to look into this rich and fascinating field.

J. B. KELLER
H. P. MCKEAN
COURANT INSTITUTE OF MATHEMATICAL SCIENCES
September, 1972

A Stochastic Problem in Visual Neurophysiology

Bruce W. Knight

1. **Introduction.** The eye of the common horseshoe crab *Limulus* may be removed from the creature, and may be maintained in the laboratory for many hours as an isolated, living visual organ. Using the isolated eye, one may insert within a single living visual cell a fine microelectrode; and thus one may observe, with the aid of sensitive amplifiers, those voltage signals which form the very basis of the horseshoe crab's visual sense. While the horseshoe crab is a particularly favorable animal for studying the electrical events which mediate between light stimulation and nervous activity, similar observations on other forms are not impossible, and experiments on numerous species of insects, spiders, and other invertebrates indicate that in the horseshoe crab we are studying a typical example of invertebrate vision. It also appears that the electrical phenomena which we observe in these visual cells arise from mechanisms which evolution has exploited frequently in other contexts throughout the nervous system. Thus it is a substantial challenge for the neurophysiologist to propose for this intracellular activity a model which, on the one hand, is in accord with known and plausible physiology, and, on the other hand, accounts for a wide range of experimental data.

These considerations have led to the proposal of the so-called "adapting bump model" by Rushton [**5**] and independently by Dodge and Adolph [**1**], [**2**]. This model of the physiology, which underlies the voltage response to light in the *Limulus* eye, satisfies the joint criteria of biological reason-

---

AMS (*MOS*) *subject classifications* (1970). Primary 60G35; Secondary 54K05, 60H10, 60K05, 60K10, 62M10, 62M15, 92-00, 92-02, 93C15, 93C20, 93C22.

ableness and of some empirical success [**3**], and has been discussed in qualitative detail elsewhere [**4**]. The present presentation will simply describe the model (first verbally and then analytically) and then will proceed to derive a few consequences: predictions suitable for an experimental check.

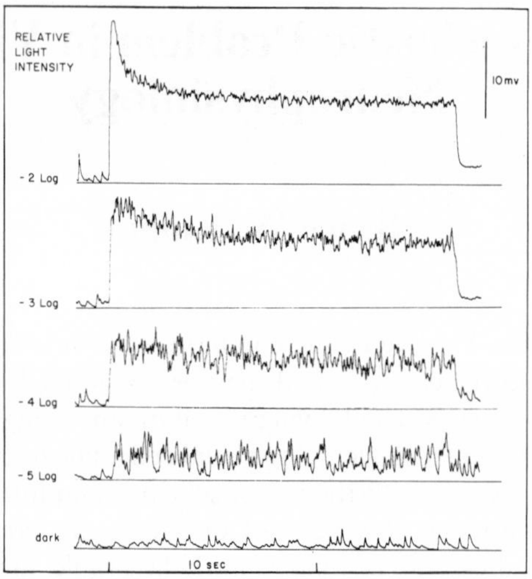

FIGURE 1. Records of the voltage within a *Limulus* visual cell in response to light of various intensities.

In dim steady light, the voltage recorded within the *Limulus* visual cell looks like a "shot noise": a superposition of voltage excursions (or "bumps") each initiated at a random point in time, and each voltage excursion in the superposition rising from zero and falling back to zero along a curve that is variable only in amplitude among shots (see Figure 1). Several sorts of evidence indicate that the continued presence of brighter light tends to shift the amplitudes of the individual shots toward smaller values. Now it is a common neurophysiological mechanism for the voltage within a cell to be changed by the sudden release of a chemical "transmitter substance" in discrete "shots" from subcellular packages or "vesicles." After a vesicle releases its shot of transmitter apparently the vesicle is recycled—it fills anew with transmitter and awaits its next release. These facts suggest the adapting bump model for the *Limulus* visual cell.

The adapting bump model assumes two things: (1) that the arrival of photons triggers the release of chemical transmitter from vesicles in the eye; and (2) that the refilling of a vesicle is a process sufficiently leisurely so that at the brighter lights a vesicle is likely to be triggered again before it has recovered its full allotment of transmitter. Thus the adapting bump model gives a plausible accounting for both the general "shot noise" character of the light-induced intracellular voltage and the apparent reduction in shot sizes at increased light intensity.

The model just qualitatively described will now be translated into dynamical equations whose consequences will be explored. Before undertaking the main work of dealing with adaptation, we dispose of the relatively much simpler matter of the "shot-shape." The voltage response of the cell to a unit impulse of transmitter may be called

(1.1)
$$g(t): \quad \int dt\, g(t) = 1, \qquad g(t < 0) = 0.$$

It has been pointed out by Fuortes and Hodgkin [**3**] that a good analytic fit to *Limulus* data may be obtained with a shot-shape of the form

(1.2)
$$g(t) = \frac{1}{\tau}\frac{(t/\tau)^n}{n!}e^{-t/\tau}$$

for $t > 0$. Leaving aside for the moment the fact that transmitter release is stochastic and in packets, if transmitter were released continuously at a rate $r(t)$, then the responding voltage $v(t)$ according to (1.1) would be given by

(1.3)
$$v(t) = \int_{-\infty}^{t} dt\, r(t')g(t - t').$$

If for formal reasons we should assume

(1.4)
$$r(t) = e^{i\omega t},$$

then

(1.5)
$$v(t) = \tilde{g}(\omega)r(t),$$

where

(1.6)
$$\tilde{g}(\omega) = \int_0^{\infty} dt\, e^{-i\omega t}g(t)$$

is the Fourier transform of (1.1). If $r(t)$ is the *expected* rate of the stochastic transmitter release, then $v(t)$ in the relations above is the expected value of the resultant stochastic voltage. If $g(t)$ is given by (1.2), then

(1.7)
$$\tilde{g}(\omega) = 1/(1 + i\tau\omega)^{n+1},$$

a form useful in explicit calculations.

2. **Adaptation.** The way the expected voltage is related to the expected

transmitter release rate has just been given. Next we formulate the relation between input light and transmitter release rate.

Call the time since a vesicle last released its transmitter, the vesicle's *age* a. Let the *size* of its content of replenished transmitter be called $S(a)$. The function $S(a)$ presumably should start at zero when $a = 0$ and the vesicle has just emptied, and increase monotonically thereafter, until it releases again. The probability of release at any age may be specified by a *hazard* $H(t, a)$ which specifies at age a the probability density, per time unit, of triggering a new release. Temporal changes in the light intensity will lead to temporal changes in the hazard at any age; thus the t dependence of $H(t, a)$ builds the effect of variable light intensity into the model.

At any time t, the population of vesicles will have a *distribution of ages* $F(t, a)$ (dimensionally a rate, $\sec^{-1}$) which, by definition, satisfies

$$\int_0^\infty da\, F(t, a) = 1. \tag{2.1}$$

Equation (2.1) means that we have chosen to normalize population variables to a per vesicle basis.

The expected total (per vesicle) *population triggering rate* $\mu(t)$ ($\sec^{-1}$) is jointly determined by hazard and population density at all ages, and is evidently

$$\mu(t) = \int_0^\infty da\, H(t, a)F(t, a). \tag{2.2}$$

Similarly the expected transmitter release rate (moles/sec, per vesicle) will be

$$r(t) = \int_0^\infty da\, S(a)H(t, a)F(t, a), \tag{2.3}$$

and it is this expression which will determine the expected voltages through equation (1.3). Since all vesicles return to age zero when triggered, the population density at age zero must equal the total triggering rate

$$F(t, 0) = \mu(t). \tag{2.4}$$

Our verbal description of the assumed dynamics of the vesicle population corresponds to the dynamical equation

$$\partial F/\partial t = -HF - \partial F/\partial a, \tag{2.5}$$

where the first right-hand term expresses attrition due to the hazard, and the second simply states that vesicles age linearly with the progress of time. The equations above give a complete mathematical description of the adapting bump model.

Equation (2.5) is easily solved by integrating along characteristic lines ($a = t + \text{const}$), and the result, with initial condition (2.4), is

$$F(t, a) = \mu(t - a) \exp\left(- \int_0^a du\, H(t - a + u, u)\right). \tag{2.6}$$

The dynamics are not yet solved at this point because μ is only known in terms of F from (2.2); in effect, (2.2) and (2.6) express a dynamical problem yet to be solved.

Before further exploring the dynamics of the general adapting bump model let us introduce two specializations which reduce it to the so-called "minimal model," which remains physiologically plausible, and which leads to some explicit analytic results without great effort. The *minimal model* assumes that young and old vesicles alike are exposed to the same hazard

$$H = H(t). \tag{2.7}$$

It secondly assumes that size as a function of age is given by a simple asymptotic exponential growth to a maximum

$$S(a) = \sigma(1 - e^{-\gamma a}), \tag{2.8}$$

where γ is the growth rate and σ is the asymptotic size.

Now, in general, the mean vesicle size will be

$$\bar{S} = \int_0^\infty da\, S(a) F(t, a). \tag{2.9}$$

According to the minimal model, H will factor from the release rate equation (2.3), giving

$$r = H\bar{S}. \tag{2.10}$$

Because of equation (2.5) it will also follow from (2.9) that

$$\frac{d\bar{S}}{dt} = - \int_0^\infty da\, S(a) \frac{\partial F(t, a)}{\partial a} - H\bar{S}. \tag{2.11}$$

By the second stipulation of the minimal model, equation (2.8) is equivalent to the differential equation

$$dS(a)/da = \gamma(\sigma - S(a)) \tag{2.12}$$

with the initial condition

$$S(0) = 0. \tag{2.13}$$

The integral in equation (2.11) may be integrated by parts, giving

$$-\int_0^\infty da\, S(a)\frac{\partial F(t,a)}{\partial a} = \int_0^\infty da \frac{dS(a)}{da} F(t,a) = \gamma(\sigma - \bar{S}(t)), \tag{2.14}$$

where (2.13) has been used in the first line, and (2.12) and (2.1) in the second. Substituting this back into (2.11) yields

$$d\bar{S}/dt = -(H(t) + \gamma)\bar{S} + \gamma\sigma. \tag{2.15}$$

An immediate consequence of equation (2.15) is that under steady light, which will give a hazard fixed at a value H_0, the mean size will relax to a value

$$\bar{S}_0 = \sigma \frac{1}{1 + H_0/\gamma} \tag{2.16}$$

exponentially with a rate $\gamma + H_0$. If the light exhibits small time-dependent departures from its mean level, then the hazard H will have a small additional time-dependent part

$$H = H_0 + \delta H(t) \tag{2.17}$$

and $\bar{S}$ likewise will respond with a small time-dependent part

$$S = \bar{S}_0 + \delta\bar{S}(t). \tag{2.18}$$

Substituting (2.17) and (2.18) into (2.15) gives the equation that the departure $\delta\bar{S}$ satisfies

$$\frac{d}{dt}\delta\bar{S} = -(H_0 + \gamma)\delta\bar{S} - \bar{S}_0\delta H. \tag{2.19}$$

(The term involving $\delta\bar{S}\delta H$ has been discarded, as it is of the second order of smallness.)

It now is easy to calculate the response of average vesicle size to a small sine-flicker superimposed on the incident light. If δH is given a time dependence of the form $e^{i\omega t}$ in (2.19), then $\delta\bar{S}$ will respond with the same time dependence, whence

$$\frac{d}{dt}\delta\bar{S} = i\omega\delta\bar{S}. \tag{2.20}$$

This result enables us to solve (2.19) for $\delta\bar{S}$, giving

$$\delta\bar{S} = -\bar{S}_0 \frac{1}{H_0 + \gamma + i\omega}\delta H. \tag{2.21}$$

From (2.10) the fluctuation in the release rate is given by

$$\delta r = \bar{S}\delta H + H\delta\bar{S}. \tag{2.22}$$

Substituting (2.21) in this relation now gives the *frequency response* (or *transfer function*) of the release rate, in response to sine-flickering light

$$\frac{\delta r}{\delta H} = \frac{\gamma\sigma}{H_0 + \gamma}\left\{1 - \frac{H_0}{H_0 + \gamma + i\omega}\right\}. \tag{2.23}$$

It is easily seen from the discussion at the end of the previous section that the corresponding frequency response of the expected voltage will be

$$\delta v/\delta H = \tilde{g}(\omega)\delta r/\delta H. \tag{2.24}$$

The reason for developing in detail an expression for the voltage frequency response is that laboratory measurements may be made, using light whose intensity level is sinusoidally modulated. Thus theory and experiment may be compared.

The most detailed experimental check of the theoretical model, a check which completely avoids the fitting of any parameters, involves a theoretical relationship, which the model predicts, between the voltage frequency response and the *autocovariance function* of the voltage signal obtained under steady light. The detailed relation is given in §3d, equation (3.30). The definition we choose for the autocovariance is the following: Let us express the voltage signal as its mean value plus fluctuations about its mean:

$$v(t) = \bar{v} + v_1(t). \tag{2.25}$$

Then the autocovariance function $c(\tau)$ is defined as

$$c(\tau) = \overline{(v_1(t)v_1(t + \tau))} \tag{2.26}$$

where the average is to be taken over all times t. Following common usage, we will refer to the Fourier transform of the autocovariance

$$\tilde{c}(\omega) = \int_{-\infty}^{\infty} dt\, e^{-i\omega\tau}c(\tau) \tag{2.27}$$

as the voltage *power spectrum*.

3. **Autocovariance for the adapting bump model.** For the sake of brevity, the following derivation of the autocovariance will be informal, including only the steps which are of strategic importance. The mathematically inclined reader will be able to supply whatever further level of rigor he finds satisfying.

a. *Autocovariance formula.* A noisy voltage signal, composed of a superposition of discrete "bump" events which have constant shape but variable size, may be expressed as

$$v(t) = \sum_n S_n g(t - t_n). \tag{3.1}$$

Here $g(t)$ is the bump shape and may be normalized in the sense that $\int g(t)\,dt = 1$. The occurrence times t_n and sizes S_n are scattered according to the statistical laws of the system which produces the bumps. The autocovariance of equation (2.26) may be re-expressed as

$$c(\tau) = -\bar{v}^2 + \lim_{T\to\infty} \frac{1}{2T} \int_{-T}^{T} dt\, v(t + \tau)v(t). \tag{3.2}$$

The second term must be evaluated, using (3.1) and the system's statistical laws.

The adapting bump model postulates a population of independent vesicles which discharge and refill independently of one another. Each vesicle will produce a voltage signal of the form (3.1). The observed voltage record will be a summation of such signals. The autocovariance of a sum of *independent* signals will be the sum of the individual autocovariances. Thus it is sufficient to determine the autocovariance resulting from the signal $v(t)$ produced by the repeated triggering of a single vesicle.

Substitution of (3.1) will introduce two summations into (3.2). The individual terms of the double sum will represent the correlation effects of individual bumps taken in all possible pairs. An alternative way of evaluating the double sum is to determine what is the expected correlation of all bumps with a single typical bump, and simply multiply that result by the number of bumps in the total time span. If events occur at a mean rate μ over the time span from $-T$ to T, their total number will be about $\mu \times 2T$. The factor $2T$ cancels the denominator in (3.2), and we conclude that the limit term in (3.2) will be μ times the expected correlation of a given bump with all others. This correlation we may examine in three pieces: a bump's correlation with its predecessors, with itself, and with its successors.

The mean correlation of a bump with itself is evidently

$$\overline{S^2} \int_{-\infty}^{\infty} dt\, g(t + \tau)g(t), \tag{3.3}$$

where the average is taken over vesicle sizes at the moment of triggering.

In our adapting bump model a vesicle carries no "memory" of its history previous to its most recent triggering. Thus knowledge that a vesicle has triggered at a given moment (say $t = 0$) is sufficient in principle to determine its expected transmitter release rate for all future times. Let us define this "conditional expected transmitter flux" $k(t)$ (moles/sec), which will be evaluated from the model in §3b below. The expected correlation of a bump with its successors and predecessors may now be expressed in terms of this conditional flux: for successors

$$\bar{S}\int_{-\infty}^{\infty} dt\, g(t+\tau)\int_{0}^{\infty} dt'\, k(t')g(t-t') \tag{3.4}$$

and for predecessors

$$\bar{S}\int_{-\infty}^{\infty} dt\, g(t+\tau)\int_{-\infty}^{0} dt'\, k(-t')g(t-t'). \tag{3.5}$$

In both expressions the first term $g(t+\tau)$ is the shape of the single bump we have selected, time-slid by an amount τ, as in (3.2). The second term (the integral) gives the correlation with other bumps, weighted by the factor k which relates bump size to time separation. The leading factor $\bar{S}$ is averaged over size at the moment of triggering and belongs to the earlier of the two bumps. As $k(t)$ is zero for negative t, the second integral may be extended between infinite limits in both expressions. The total autocovariance may thus be written as

$$\begin{aligned} c(\tau) = -\bar{v}^2 + \mu\int_{-\infty}^{\infty} dt\, g(t+\tau) \\ \cdot\left\{g(t)\overline{S^2} + \bar{S}\int_{-\infty}^{\infty} dt'(k(t') + k(-t'))g(t-t')\right\}. \end{aligned} \tag{3.6}$$

For long elapsed times, the conditional transmitter flux $k(t)$ must eventually go to its unconditional value $\mu\bar{S} = \bar{v}$. Because g is normalized, it may be seen that the long expression in (3.6) goes to $\bar{v}^2$ for large $|\tau|$, giving $c(\tau) \to 0$.

b. *Conditional flux.* In order to evaluate (3.6), all that remains is to determine the conditional flux $k(t)$, and this is a problem of only a technical nature. Define the "triggering density"

$$f(a) = (1/\mu)H(a)F(a). \tag{3.7}$$

(Here μ and $F(a)$ are as defined in §2. As the hazard is steady in steady light, there is no time dependence.) The two averages in (3.6) will be given by

$$\overline{S^n} = \int_{0}^{\infty} da(S(a))^n f(a). \tag{3.8}$$

(Notice that this average is weighted according to the density of *triggering* vesicles, and will be different from the average of the previous section, if the hazard is age-dependent. However, the result of that section was specific to the minimal model, for which (3.8) reduces to the time-independent form of (2.9).) Define also the contribution to the conditional transmitter flux furnished by the first subsequent triggering event

$$b(t) \equiv S(t)f(t). \tag{3.9}$$

The total conditional flux will be this term plus the contribution of all later events:

$$k(t) = b(t) + \int_0^t dt'\, f(t')k(t - t'). \tag{3.10}$$

This is an integral equation to be solved for $k(t)$.

Since both $f(t)$ and $k(t)$ are zero for negative t, the integration limits in (3.10) may be extended to $\pm\infty$, and the Fourier convolution theorem may be applied. This yields

$$\tilde{k}(\omega) = \tilde{b}(\omega)/(1 - \tilde{f}(\omega)), \tag{3.11}$$

where

$$\tilde{k}(\omega) = \int_{-\infty}^{\infty} dt\, e^{-i\omega t}k(t), \quad \text{etc.,} \tag{3.12}$$

whence formally

$$k(t) = \frac{1}{2\pi}\int_{-\infty}^{\infty} d\omega\, e^{i\omega t}\frac{\tilde{b}(\omega)}{1 - \tilde{f}(\omega)}. \tag{3.13}$$

Here an important and interesting technical complexity arises. Because $f(t)$ is normalized, $\tilde{f}(0) = 1$, and the integrand of (3.13) is singular at $\omega = 0$. (The difficulty is already implicit in (3.12), as $\tilde{k}(t)$ should go to a nonzero constant at large t, and ω must be given a small negative imaginary part to make the integral converge.) However, $\tilde{k}$ may be written in the form

$$\tilde{k}(\omega) = \tilde{n}(\omega) + A/i\omega, \tag{3.14}$$

where $\tilde{n}(\omega)$ is nonsingular. The corresponding time function is

$$k(t) = n(t)\begin{cases} -A & \text{if } t > 0, \\ +0 & \text{if } t < 0, \end{cases} \tag{3.15}$$

where $n(t)$ goes to zero for large t. The value of A will be given by the first Taylor-series coefficient of $(\tilde{k}(\omega))^{-1}$:

$$A = i\left(\frac{d}{d\omega}(\tilde{k}(\omega))^{-1}_{\omega=0}\right)^{-1} = \frac{i\tilde{b}(0)}{(-d\tilde{f}/d\omega)_{\omega=0}}. \tag{3.16}$$

Now $\tilde{b}(0) = \bar{S}$, while

$$\left(\frac{d\tilde{f}}{d\omega}\right)_{\omega=0} = \int_{-\infty}^{\infty} dt\,(-itf(t)) = -i\bar{t} = -i\mu^{-1}. \tag{3.17}$$

Thus

$$A = \mu \bar{S} = \bar{v} \tag{3.18}$$

and $k(t \to +\infty) = \bar{v}$. It is the singularity in $\tilde{k}(\omega)$ that cancels the $-\bar{v}^2$ term in (3.6), which becomes

$$c(\tau) = \mu \int_{-\infty}^{\infty} dt\, g(t+\tau)\left\{ g(t)\overline{S^2} + \bar{S} \int_{-\infty}^{\infty} dt'\, g(t-t')(n(t)+n(-t)) \right\}, \tag{3.19}$$

where now $\overline{S^2}$, $\bar{S}$ and n are all known. Using the convolution theorem on (3.19) gives a particularly simple expression for the power spectrum

$$\begin{aligned} \tilde{c}(\omega) &= \mu \tilde{g}(\omega)\tilde{g}(-\omega)\{\overline{S^2} + \bar{S}(\tilde{n}(\omega) + \tilde{n}(-\omega))\} \\ &= \mu \tilde{g}(\omega)\tilde{g}(-\omega)\{\overline{S^2} + \bar{S}(\tilde{k}(\omega) + \tilde{k}(-\omega))\}. \end{aligned} \tag{3.20}$$

Since $\tilde{k}(\omega)$ is directly calculated in (3.11), the convenient way to express the autocovariance is

$$c(\tau) = \mu \frac{1}{2\pi} \int_{-\infty}^{\infty} d\omega\, e^{i\omega t} \tilde{g}(\omega)\tilde{g}(-\omega)\{\overline{S^2} + \bar{S}(\tilde{k}(\omega) + \tilde{k}(-\omega))\}. \tag{3.21}$$

c. *Minimal model.* In the minimal model all the pieces of (3.20) are straightforward to evaluate:

$$\mu = H_0, \tag{3.22}$$

$$\bar{S} = \sigma \frac{\gamma}{H_0 + \gamma}, \tag{3.23}$$

$$\overline{S^2} = \sigma \frac{2\gamma^2}{(H_0+\gamma)(H_0+2\gamma)}, \tag{3.24}$$

$$\tilde{b}(\omega) = \sigma H_0 \left\{ \frac{1}{H_0 + i\omega} - \frac{1}{H_0 + \gamma + i\omega} \right\}, \tag{3.25}$$

$$\tilde{f}(\omega) = \frac{H_0}{H_0 + i\omega} = 1 - \frac{i\omega}{H_0 + i\omega}, \tag{3.26}$$

$$\tilde{k}(\omega) = \frac{\sigma\gamma H_0}{H_0+\gamma}\left(\frac{1}{i\omega} - \frac{1}{H_0+\gamma+i\omega} \right), \tag{3.27}$$

$$\begin{aligned} \tilde{c}(\omega) &= \mu |\tilde{g}(\omega)|^2 (\sigma\gamma)^2 \frac{2}{(H_0+\gamma)(H_0+2\gamma)} \\ &\quad \cdot \left\{ 1 - \frac{H_0(H_0+2\gamma)}{2(H_0+\gamma)} \left(\frac{1}{H_0+\gamma+i\omega} + \frac{1}{H_0+\gamma-i\omega} \right) \right\}. \end{aligned} \tag{3.28}$$

Here $\tilde{g}(\omega)$ is given by equation (1.7). The Fourier transform which evaluates $c(\tau)$ may be done analytically by the method of residues.

d. *"Nonparametric" relation with frequency response.* If equation (2.23) is multiplied by its complex conjugate, and separated into partial fractions, the result is

$$\left|\frac{\delta r}{\delta H}\right|^2 = (\sigma\gamma)^2 \frac{1}{(H_0+\gamma)^2} \cdot \left\{1 - \frac{H_0(H_0+2\gamma)}{2(H_0+\gamma)}\left(\frac{1}{H_0+\gamma+i\omega} + \frac{1}{H_0+\gamma-i\omega}\right)\right\}. \tag{3.29}$$

Multiplying by the absolute square of the normalized bump transfer function, and comparison with (3.28) gives

$$\tilde{c}(\omega) = \frac{2H_0(H_0+\gamma)}{H_0+2\gamma}\left|\tilde{g}(\omega)\frac{\delta r}{\delta H}\right|^2. \tag{3.30}$$

Hence, the power spectrum and squared frequency response are proportional. This implies that the experimentally measured voltage frequency response (2.24) may be used (by substituting (3.30) into (2.27) and Fourier transforming) to predict the autocovariance. The prediction and direct measurement are compared in Figure 2.

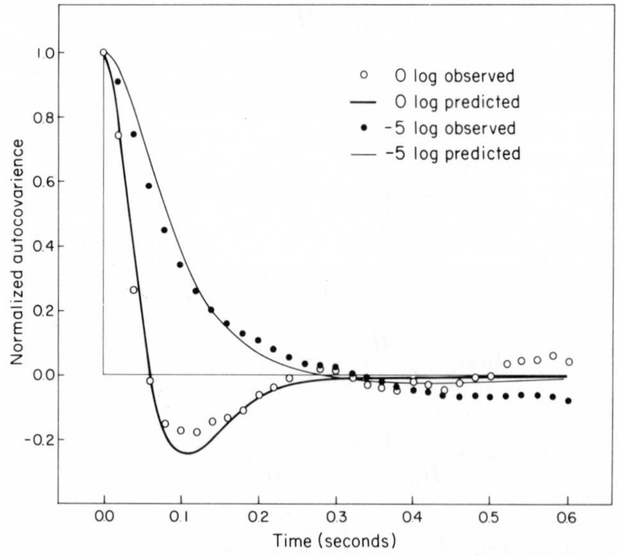

FIGURE 2. Autocovariance of the intracellular voltage fluctuations in response to steady light. Points show direct measurements. Lines show prediction from the frequency response, measured on the same cell with flickering light. The bright light (0 log) was 10^5 times as intense as the dim light (-5 log).

e. *Generalized Campbell's theorem.* Algebraic rearrangement will bring (3.28) to the form

$$\tilde{c}(\omega) = \mu \bar{S}^2 \frac{2\gamma^2}{(H_0 + 2\gamma)(H_0 + \gamma)} |T(\omega)|^2, \tag{3.31}$$

where $T(\omega)$ is the normalized transfer function (2.24) for the full flicker process. If there are N vesicles per cell, the total power spectrum will be given by

$$\tilde{c}_T(\omega) = N\tilde{c}(\omega). \tag{3.32}$$

The variance of the voltage signal is the autocovariance for $\tau = 0$, whence

$$\text{Var} = N\mu \bar{S}^2 \frac{2\gamma^2}{(H_0 + 2\gamma)(H_0 + \gamma)} \frac{1}{2\pi} \int_{-\infty}^{\infty} d\omega\, |T(\omega)|^2. \tag{3.33}$$

The integral has already been evaluated in the Fourier inversion of (3.28); the variance is directly measurable from laboratory observations; so the only unknowns in (3.33) are N and $\bar{S}$. Now the mean voltage is given by

$$\bar{v} = N\mu\bar{S} \tag{3.34}$$

and since $\bar{v}$ is also directly observable, the only unknowns again are N and $\bar{S}$, which may be solved for simultaneously from (3.33) and (3.34).

4. **Distribution of sizes.** The probability density over bump sizes, $P(s)$, will satisfy

$$P(s)\, ds = F(a)\, da \quad \text{or} \tag{4.1}$$

$$P = F(a)/(ds/da). \tag{4.2}$$

For the minimal model this evaluates to

$$P = (H_0/\sigma\gamma)e^{-(H_0 - \gamma)a}. \tag{4.3}$$

Solving equation (2.8) for a gives

$$a = -(1/\gamma)\ln(1 - s/\sigma) \tag{4.4}$$

and substitution into (4.3) gives the probability density for sizes:

$$P(s) = \frac{H_0}{\sigma\gamma}\left(1 - \frac{s}{\sigma}\right)^{(H_0/\gamma - 1)} \tag{4.5}$$

For hazard smaller than growth rate, the distribution is sharply skewed toward maximum size. This ceases to be true when $H_0 = \gamma$. When $H_0 > 2\gamma$, sizes near maximum become extremely rare. Nonetheless such anomalously large bumps should be seen on infrequent occasions.

5. **General adaptation transfer function.** In §2 the adaptation frequency response for the minimal model was derived from a simple differential equation. This efficient approach is not possible for more general assumptions. The more laborious general method is presented below.

For a time-independent hazard $H_0(a)$ the resulting age distribution will be

$$F_0(a) = \mu_0 \exp\left(- \int_0^a du\, H_0(u) \right), \tag{5.1}$$

which follows from specializing (2.6). Assume now a small fluctuating perturbation superimposed on the steady state of each variable:

$$H(t, a) = H_0(a) + H_1(a)e^{i\omega t}, \tag{5.2}$$

$$F(t, a) = F_0(a) + F_1(a)e^{i\omega t}, \tag{5.3}$$

$$\mu(t) = \mu_0 + \mu_1 e^{i\omega t}, \tag{5.4}$$

$$r(t) = r_0 + r_1 e^{i\omega t}. \tag{5.5}$$

Here $H_1(a)$ is assumed to be given (by model assumptions concerning the transduction from light intensity to hazard) but the other perturbation variables must be determined from the dynamics of the adaptation model.

The dynamic equations of adaptation are (2.2), (2.3), and (2.6). The former two yield at once the perturbation equations

$$\mu_1 = \int_0^\infty da\, (F_0(a)H_1(a) + H_0(a)F_1(a)), \tag{5.6}$$

$$r_1 = \int_0^\infty da\, S(a)(F_0(a)H_1(a) + H_0(a)F_1(a)). \tag{5.7}$$

(Notice that specializing (5.7) to the minimal model yields equation (2.22).) Equation (2.6) yields the perturbation equation

$$\begin{aligned} F_1(a) &= \mu_1 e^{-i\omega a} \exp\left(- \int_0^a du\, H_0(u) \right) \\ &\quad + \mu_0 \left\{ - e^{-i\omega a} \int_0^a du\, e^{i\omega u} H_1(u) \right\} \exp\left(- \int_0^a du\, H_0(u) \right). \end{aligned} \tag{5.8}$$

Substituting (5.8) into (5.6) now yields

$$\mu_1 = \mu_0 \frac{\displaystyle\int_0^\infty da \exp\left(- \int_0^a du\, H_0(u) \right) \left\{ H_1(a) - e^{-i\omega a} H_0(a) \int_0^a du\, e^{i\omega u} H_1(u) \right\}}{\displaystyle 1 - \int_0^\infty da\, e^{-i\omega a} H_0(a) \exp\left(- \int_0^a du\, H_0(u) \right)}. \tag{5.9}$$

Substitution back into (5.8) in turn determines $F_1(a)$. We note that the denominator of (5.9) is $1 - \tilde{f}(\omega)$, the denominator of $\tilde{k}(\omega)$ in (3.11). Finally substitute $F_1(a)$ into (5.7) to obtain

$$r_1 = \mu_0 \int_0^\infty da \exp\left(-\int_0^a du\, H_0(u)\right)(S(a) + \tilde{k}(\omega)) \cdot \left[H_1(a) - e^{-i\omega a} H_0(a) \int_0^a du\, e^{i\omega u} H_1(u)\right], \tag{5.10}$$

whose evaluation is now a matter of performing the several integrations. Any model in which the hazard has the form

$$H(t, a) = \Phi(t)\Psi(a) \tag{5.11}$$

will yield a hazard perturbation of the form

$$H_1(a) = \varepsilon H_0(a). \tag{5.12}$$

The adaptation transfer function in this case simplifies a bit to

$$\frac{r_1}{\varepsilon} = \mu_0 \int_0^\infty da\, f(a)(S(a) + \tilde{k}(\omega))\left[1 - e^{-i\omega a}\int_0^a du\, e^{i\omega u} H_0(u)\right], \tag{5.13}$$

where $f(a)$ was defined in equation (3.7). The minimal model and "dead time" model fall into this category.

It must be mentioned here that the remarkable relationship (3.30) of the form

$$\tilde{c}(\omega) = A|\tilde{g}(\omega)\delta r/\delta H|^2, \tag{5.14}$$

where A is a constant, which led to the nonparametric comparison between theory and experiment shown in Figure 2, does *not* follow from the general adapting bump model results presented in this section. We make the following observations:

(a) The *minimal model without growth*

$$H = H(t),$$
$$S(a) = \text{const}$$

satisfies (5.14). It yields ordinary Poisson shot-noise, which satisfies the relationship in a fairly trivial way. This is a limiting case of the minimal model with $\gamma \to \infty$.

(b) The *minimal model with linear growth*

$$H = H(t),$$
$$S(a) = S(1)a$$

satisfies (5.14). It is easy to show that this is also a limiting case ($\gamma \to 0$, $\sigma \sim 1/\gamma$).

(c) The *dead time model*

$$\begin{aligned} H(t, a) &= 0 && \text{for } a < a_1, \\ &= H(t) && \text{for } a > a_1, \\ S(a) &= \text{const}, \end{aligned}$$

satisfies (5.14), even though the hazard is age-dependent in this case.

(d) The *Limulus eye* satisfies (5.14) rather closely over a wide range of light intensities, as Figure 2 shows.

(e) The model defined by

$$H = H(t), \qquad S(a) = \sigma\{1 - \alpha e^{-\gamma_1 a} - (1 - \alpha)e^{-\gamma_2 a}\}$$

does *not satisfy* (5.14) in general, even though it reduces to the minimal model if $\alpha = 1$ or if $\alpha = 0$ or if $\gamma_1 = \gamma_2$. Thus we are left with an unsolved challenge: To define the just necessary additional conditions such that the adapting bump model will satisfy (5.14), exactly or to within experimental accuracy.

6. **Green's function formulation.** It may be that several unsolved problems, for example the range of validity of the nonparametric procedure, discussed just above, will be brought closer to solution by the introduction of further mathematical tools. Thus we remark here that the frequency response and autocovariance can be derived from a single unified approach: The use of a "propagator" or "Green's function" formulation. The general method is well known; its application to the adapting bump model will be presented without detailed elucidation.

The boundary condition (2.4) and partial differential equation (2.5) may be combined into the single nonlocal linear relationship

$$\frac{\partial F}{\partial t} = -\frac{\partial F}{\partial a} - HF + \delta(a)\int_0^\infty da'\, H(t, a')F(t, a'). \tag{6.1}$$

By defining the linear operator E through the property

$$Eu(a) \equiv \delta(a)\int_0^\infty da'\, u(a'), \tag{6.2}$$

equation (6.1) becomes

$$\partial F/\partial t = -\partial F/\partial a - (1 - E)HF \equiv AF. \tag{6.3}$$

Now E is a projective operator in the sense that it linearly maps all functions of a to the particular function $\delta(a)$, and has several useful simple properties, such as

$$EQE = \phi E \tag{6.4}$$

where Q is an arbitrary operator and ϕ is a scalar.

If $H = H_0$ is independent of time, then (6.3) has the simple formal solution

$$\begin{aligned} F(t) &= e^{At}F(0) \quad \text{for } t > 0, \\ &= 0 \qquad\qquad \text{for } t < 0 \end{aligned} \tag{6.5}$$

(here age dependence remains implicit). Thus determining the evolution of the age distribution in time is equivalent to obtaining an explicit expression for the operator

$$e^{At} = G_t, \tag{6.6}$$

the propagator or Green's function of the system. Taking the Fourier transform of (6.6) yields the frequency spectrum of G_t:

$$\tilde{G}(\omega) = (i\omega - A)^{-1}. \tag{6.7}$$

The problem is thus reduced to finding the inverse operator to

$$i\omega - A = i\omega + \partial/\partial a + H - EH \equiv B - EH. \tag{6.8}$$

Here B is a simple differential operator, and its inverse

$$\Gamma = B^{-1} = (\partial/\partial a + H(a) + i\omega)^{-1} \tag{6.9}$$

is easily found. Then the full inverse operator to (6.8) is given by

$$\tilde{G} = \left(1 + \frac{1}{1 - \tilde{f}} \Gamma EH\right) \Gamma \tag{6.10}$$

where the scalar $\tilde{f}(\omega)$ in turn is given by

$$EH\Gamma E = \tilde{f}E. \tag{6.11}$$

Proof is by multiplying (6.10) and (6.8). Evaluation in detail will show that the function $\tilde{f}(\omega)$ is the same as was encountered in the denominators of (3.11) and (5.9). The resulting singularity in $\tilde{G}_\omega$ at $\omega = 0$ may be traced to the fact that there is an equilibrium distribution $F_0(a)$ which in (6.3) will give

$$AF_0 = 0. \tag{6.12}$$

The corresponding Fourier integral of (6.5) will diverge at $\omega = 0$.

We now show that both the adaptation frequency response and the autocovariance follow from the Green's function. Augmenting the time-independent operator A of (6.3) by a perturbation of the form

$$A_1 e^{i\omega t} = (1 - E)H_1 e^{i\omega t} \tag{6.13}$$

yields the perturbation equation

$$i\omega F_1 = A_0 F_1 - (1 - E)H_1 F_0, \tag{6.14}$$

whence

$$F_1 = -\tilde{G}_\omega (1 - E)H_1 F_0. \tag{6.15}$$

Substituting into the transmitter release equation (5.7) then gives

$$r_1 = \int_0^\infty da\, S(a)\left\{1 - H_0 \tilde{G}_\omega (1 - E)\right\} H_1 F_0. \tag{6.16}$$

The operator E corresponds to vesicle triggering and return to age zero. The operator G_t then gives the resulting probability distribution at a later time t. The correlation in transmitter release between two distinct triggerings must be weighted by both vesicle size, and triggering probability or hazard, and will be given by

$$c'(t) = \int_0^\infty da\, SHG_{|t|}ESHF_0. \tag{6.17}$$

The prime is to remind us that both the correlation of a triggering event with itself, and the effects of finite bump duration have been left out. The absolute bars make the expression applicable to either positive or negative t. Because of (6.5),

$$G_{|t|} = G_t + G_{-t} \tag{6.18}$$

and $c'(t)$ yields the power spectrum

$$\begin{aligned} \tilde{c}'(\omega) &= \int_0^\infty da\, SH(\tilde{G}_\omega + \tilde{G}_{-\omega})ESHF_0 \\ &= (\tilde{k}(\omega) + \tilde{k}(-\omega))\bar{S}, \end{aligned} \tag{6.19}$$

the last line following when the evaluations are gone through in detail. Note that (6.19) is identical to the most difficult part of (3.20). The Green's function method tends to systematize and simplify matters of routine evaluation somewhat. However, its foremost virtue is that it removes purely algebraic complexities to the background until they are called for, and leaves both the physical and the formal features of the system more plainly in sight.

REFERENCES

1. A. R. Adolph, *Spontaneous slow potential fluctuations in the Limulus photoreceptor*, J. Gen. Physiol. **48** (1964), 297–322.

2. F. A. Dodge, Jr., B. W. Knight, Jr., and J. Toyoda, *Voltage noise in Limulus visual cells*, Science **1968**, 88–90.

3. M. G. F. Fuortes and A. L. Hodgkin, *Changes in the time scale and sensitivity in the ommatidia of Limulus*, J. Physiol. **172** (1964), 239–263.

4. B. W. Knight, Jr., "Some point processes in motor and sensory neurophysiology," in *Stochastic point processes*, P. A. W. Lewis (editor), Wiley, New York, 1972. (In this reference beware of four errors in Figures 2, 8, 22 or their captions.)

5. W. A. H. Rushton, "The intensity factor in vision," in *Light and life*, W. D. McElroy and B. Glass (editors), John Hopkins Press, Baltimore, Md., 1961, pp. 706–722.

ROCKEFELLER UNIVERSITY

Some Problems and Methods for the Analysis of Stochastic Equations

George C. Papanicolaou

1. **Introduction.** Stochastic equations arise in a great variety of problems and in a very natural and intuitively clear way. Their analysis is, however, difficult, much more than the analysis of their deterministic counterparts. We shall present here a class of problems which admit a very satisfactory solution in a certain asymptotic limit. This limit corresponds to and is a generalization of the classical central limit theorem.

In §2 we state the problems with which we are concerned. Then we give three examples arising in the theory of oscillations and in one-dimensional wave propagation. Next we state a theorem which characterizes the asymptotic behavior of the processes defined by the differential equations we consider. Finally we apply this theorem to the examples. There are, of course, many other problems of interest which we do not consider here. Some of these are considered in the article of Morrison and McKenna [**1**].

In §3 we introduce certain operator valued processes called random evolutions. An extensive review of work on random evolutions and their asymptotic properties with many interesting remarks and applications can be found in [**2**]. Here we give only a brief survey of some asymptotic results and point out the fact that random evolutions are naturally associated with stochastic differential equations while their asymptotic properties lead to results on singular perturbations of partial differential equations.

2. **Initial value problems for stochastic ordinary differential equations.**

2.1. **Statement of the problem.** Let $x(t)$, $t \geqq 0$, be a stochastic process

AMS (MOS) subject classifications (1970). Primary 60-02; Secondary 60H10, 34F05.

with values in R^m and define the process $z(t) \in R^n$ as the solution of

$$dz(t)/dt = F(z(t), x(t), t), \qquad z(0) = z. \tag{2.1.1}$$

We are concerned with the following problem. Given the statistical characteristics of $x(t)$, determine the statistical characteristics of $z(t)$.

Without any assumptions on the process $x(t)$, which we may conceive as the coefficients or external influences in equation (2.1.1), and without any assumptions on the function $F(z, x, t)$, beyond some smoothness to ensure existence and uniqueness, we can not hope to obtain much information about $z(t)$. Let us consider some special types of problems.

Consider first the equation

$$dz(t)/dt = A(t)z(t) + Bx(t), \qquad z(0) = z. \tag{2.1.2}$$

Here $A(t)$ is an $n \times n$ (nonrandom) matrix and B is an $n \times m$ constant matrix. Let $Y(t)$ denote the solution matrix of (2.1.2) with $x \equiv 0$. Then the solution of (2.1.2) is easily obtained and from it all statistical characteristics of $z(t)$ can be computed directly.

Consider next the problem

$$dz(t)/dt = f(z(t)) + Bx(t), \qquad z(0) = z. \tag{2.1.3}$$

Here f is a differentiable function of R^n into R^n. This problem is considerably more involved than (2.1.2). As it stands, little can be said about $z(t)$. Suppose however that the expected value of $x(t)$, $E\{x(t)\}$, is zero and that there is a small parameter ε in the problem so that

$$dz(t)/dt = \varepsilon^2 f(z(t)) + \varepsilon Bx(t), \qquad z(0) = z. \tag{2.1.4}$$

Then the statistical properties of $z(t)$ can be obtained in the limit $\varepsilon \to 0$, t of order $1/\varepsilon^2$, as we shall see in §2.3. It is frequently assumed, also, that $x(t)$ is a white noise process, in which case (2.1.3) is interpreted as a Langevin equation for the process $z(t)$ [**3**], [**4**].

The problem of primary interest to us is the following. We assume that there is a small parameter ε and that

$$dz(t)/dt = \varepsilon F(z(t), x(t), t), \qquad z(0) = z. \tag{2.1.5}$$

We assume further that, for fixed z,

$$E\{F(z, x(t), t)\} = 0. \tag{2.1.6}$$

Then we consider the process $z(t)$ when ε is small, t is of order $1/\varepsilon^2$ and $x(t)$ is a process whose correlation length is finite. It is this problem that gives rise to processes $z(t)$ in which stochastic effects are most prominent.

Many physical problems lead to equations of the form (2.1.5) [**5**]. If no small parameter ε is present one may take $x(t)$ to be Markovian in which case $z(t)$ and $x(t)$ constitute jointly a Markov process. It is then possible to obtain information about $z(t)$ by studying the appropriate Fokker-Planck-Kolmogorov equation [**5**], [**6**]. For dynamical systems with few degrees of freedom, the equilibrium solutions of the Fokker-Planck-Kolmogorov equation may frequently be constructed even when the full equation can not be solved [**5**].

If $x(t)$ in (2.1.5) is white noise we may interpret (2.1.5) as an Itô equation [**7**] and proceed directly with the study of the associated diffusion process. Some caution must be exercised when doing this because coordinate invariance of the original dynamical system may be violated. This comes about from the well-known difficulties associated with the white noise [**8**].

Questions of "stability," of various types, for the process $z(t)$ can be studied either by methods analogous to the ones for deterministic equations [**9**], [**10**], or by considering $z(t)$ to be a diffusion Markov process and analyzing it [**9**].

One reason for considering (2.1.5), (2.1.6) in the appropriate asymptotic limit, rather than proceeding along a different route is that, for many problems, such as those of wave propagation, §2.2, this set up seems most natural. Also, the characteristics of $z(t)$ frequently acquire, in the asymptotic limit, a particularly simple form which is very desirable, especially when the problem has many degrees of freedom.

2.2. **Some examples.** We shall consider three basic examples of the types of problems to which the asymptotic analysis of §2.3 applies. The first two are oscillation problems and the last a one-dimensional wave propagation problem.

Let $z(t) \in R^1$ be the process defined as the solution of

$$\begin{aligned} \ddot{z} + \omega^2 z &= \varepsilon^2 h(z, \dot{z}) + \varepsilon x(t), \\ z(0) &= z_0, \\ \dot{z}(0) &= \dot{z}_0. \end{aligned} \tag{2.2.1}$$

Here the dot stands for time derivative, ε is a small parameter, $x(t) \in R^1$ is a zero mean stationary stochastic process and $h(z, \dot{z})$ is a smooth function on R^2. When $x(t) \equiv 0$, (2.2.1) is the basic problem first studied extensively by Krylov and Bogoljubov [**11**]. The stochastic version (2.2.1) was treated in detail by Stratonovič [**5**]. To bring this problem into the form (2.1.5) we introduce the van der Pol variables $r(t)$ and $\phi(t)$ by

$$\begin{aligned} z &= r\cos(\omega t + \phi), \\ \dot{z} &= r\sin(\omega t + \phi). \end{aligned} \tag{2.2.2}$$

From (2.2.1) and (2.2.2), we find that r and ϕ satisfy the equations

$$
\begin{aligned}
\dot{r} &= -\frac{\varepsilon}{\omega}x(t)\sin(\omega t + \phi) \\
&\quad - \frac{\varepsilon^2}{\omega}h(r\cos(\omega t + \phi), -\omega r\sin(\omega t + \phi))\sin(\omega t + \phi), \\
\dot{\phi} &= -\frac{\varepsilon}{\omega}x(t)\cos(\omega t + \phi) \\
&\quad - \frac{\varepsilon^2}{\omega}h(r\cos(\omega t + \phi), -\omega r\sin(\omega t + \phi))\cos(\omega t + \phi).
\end{aligned} \tag{2.2.3}
$$

These equations are of the form (2.1.5) with an additional nonstochastic term of order ε^2 on the right side. The presence of this term presents no additional difficulties.

Let us explain the role of the parameter ε. When $x(t) = 0$ in the above problem the Krylov-Bogoljubov [**11**] analysis says that $r(t)$, $\phi(t)$ are well approximated by the solutions of another set of equations, the averaged version of (2.2.3), for times up to order $1/\varepsilon^2$. Similarly, when $h = 0$ the solution of (2.2.1) is obtained easily by integration and it involves, essentially, $\varepsilon\int_0^t x(s)\,ds$ which behaves like a Gaussian random variable for ε small, $t \sim 1/\varepsilon^2$, assuming that the classical central limit theorem holds [**13**]. Thus the scaling in (2.2.1), (2.2.3) permits the simultaneous consideration of nonlinear and stochastic effects.

Consider next the process defined by

$$
\ddot{z} + \omega^2[1 + \varepsilon x(t) + \varepsilon^2\alpha\sin\beta t]z = 0, \qquad z(0) = z_0, \qquad \dot{z}(0) = \dot{z}_0. \tag{2.2.4}
$$

Here α and $\beta > 0$ are constants. Note that (2.2.4) is a linear problem but the coefficients are variable and contain a stochastic process $x(t)$. This makes the analysis of (2.2.4) about as difficult as that of (2.2.1). Problem (2.2.4) was first considered by Stratonovič [**5**]. We may again transform (2.2.4) by using (2.2.2) where instead of r we put e^r. This yields

$$
\begin{aligned}
\dot{r} &= \frac{\varepsilon\omega x(t)}{2}\sin(2\omega t + 2\phi) + \frac{\varepsilon^2\omega\alpha}{2}\sin\beta t\sin(2\omega t + 2\phi), \\
\dot{\phi} &= \frac{\varepsilon\omega x(t)}{2}[1 + \cos(2\omega t + 2\phi)] + \frac{\varepsilon^2\omega\alpha}{2}\sin\beta t[1 + \cos(2\omega t + 2\phi)],
\end{aligned} \tag{2.2.5}
$$

which is in the form (2.1.5).

Let us now consider a one-dimensional wave propagation problem. Let $t \in (-\infty, \infty)$ denote the space variable and $z(t)$ the time-harmonic complex valued wave field at location t. Then $z(t)$ satisfies the equation

$$
\ddot{z} + k^2n^2(t)z = 0. \tag{2.2.6}
$$

Here k is the free space wave number and $n(t)$ is the index of refraction. We assume that

$$n^2(t) = 1 + \varepsilon x(t), \tag{2.2.7}$$

where $x(t)$ is a real zero mean stationary stochastic process. We now represent the field $z(t)$ in the form

$$\begin{aligned} z(t) &= k^{1/2}[e^{ikt}a(t) + e^{-ikt}b(t)], \\ \dot{z}(t) &= ik^{3/2}[e^{ikt}a(t) - e^{-ikt}b(t)], \qquad i = (-1)^{1/2}. \end{aligned} \tag{2.2.8}$$

Upon using (2.2.6), (2.2.7) and (2.2.8) we obtain the following equations for the complex amplitudes $a(t)$ and $b(t)$:

$$\begin{pmatrix} \dot{a} \\ \dot{b} \end{pmatrix} = \varepsilon k x(t) \begin{pmatrix} e^{-ikt} & 0 \\ 0 & e^{ikt} \end{pmatrix} \begin{pmatrix} i/2 & i/2 \\ -i/2 & -i/2 \end{pmatrix} \begin{pmatrix} e^{ikt} & 0 \\ 0 & e^{-ikt} \end{pmatrix} \begin{pmatrix} a \\ b \end{pmatrix}. \tag{2.2.9}$$

The quantities $a(t)$ and $b(t)$ are the amplitudes of the forward and backward waves which make up $z(t)$. Clearly (2.2.9) is of the form (2.1.5) except that, because of the interpretation of a and b, we do not have an initial value problem. To obtain an initial value problem we proceed as follows. Let the random medium occupy the interval $[0, L]$. We prescribe $a(0) = 1$ and $b(L) = 0$ which means that waves of unit amplitude are incident from the left. Then $b(0) = R$ is the reflection coefficient and $a(L) = T$ is the transmission coefficient. Let $Y(t)$ be the fundamental solution matrix of (2.2.9), $Y(0) = I$ the identity. We have

$$\begin{pmatrix} a(L) \\ b(L) \end{pmatrix} = \begin{pmatrix} Y_{11}(L) & Y_{12}(L) \\ Y_{21}(L) & Y_{22}(L) \end{pmatrix} \begin{pmatrix} a(0) \\ b(0) \end{pmatrix}, \tag{2.2.10}$$

and hence

$$R(L) = -Y_{21}(L)/Y_{22}(L), \qquad T(L) = 1/Y_{22}(L). \tag{2.2.11}$$

Here we have used the fact that $\det Y(t) = 1, t \geqq 0$. From (2.2.11) it follows that the statistical properties of $R(L)$ and $T(L)$ can be obtained from those of $Y(L)$, which is a matrix-valued solution of (2.2.9) with $Y(0) = I$. This is an initial value problem, but for a matrix-valued random process.

The final step in bringing the problem for $Y(t)$ into the form (2.1.5) is to introduce variables analogous to the van der Pol variables. This is accomplished by setting

$$Y(t) = \begin{pmatrix} e^{i\phi/2} & 0 \\ 0 & e^{-i\phi/2} \end{pmatrix} \begin{pmatrix} \cosh \theta/2 & \sinh \theta/2 \\ \sinh \theta/2 & \cosh \theta/2 \end{pmatrix} \begin{pmatrix} e^{i\psi/2} & 0 \\ 0 & e^{-i\psi/2} \end{pmatrix}, \tag{2.2.12}$$

where $\phi = \phi(t), \theta = \theta(t), \psi = \psi(t)$.

In these variables (2.2.9) becomes

$$\begin{aligned} \dot{\phi} &= \varepsilon k x(t)[1 + \coth \theta \cos(2kt + \phi)], \qquad \dot{\theta} = \varepsilon k x(t) \sin(2kt + \phi), \\ \dot{\psi} &= -\varepsilon k x(t) \operatorname{cosech} \theta \cos(2kt + \phi). \end{aligned} \tag{2.2.13}$$

Now we have a problem of the type (2.1.5).

Let us consider briefly (2.2.13). First we note that θ and ϕ decouple from ψ. This has the following consequence. If we define R and T when $b(L) = 1$ and $a(0) = 0$, i.e., for waves incident from the right, then

$$R(L) = Y_{12}(L)/Y_{22}(L) = e^{i\phi} \tanh \theta/2. \tag{2.2.14}$$

Thus the reflection coefficient from the right decouples from the rest of Y and may be considered by itself [**28**]. On the other hand, the reflection coefficient from the left given by (2.2.11) (and equal to $-e^{i\psi} \tanh \theta/2$), even though it does not decouple at this stage, behaves better in the asymptotic limit of the next section.

The analysis of the wave propagation problem is due to R. Burridge and the author. The above results and the asymptotic analysis which follows have been generalized to the coupled mode problem where $z(t)$ in (2.2.6) is a complex n-vector, k^2 a real diagonal $n \times n$ matrix and $n^2(t)$ a real, symmetric, $n \times n$ matrix-valued stochastic process [**14**].

2.3. **Asymptotic analysis.** We now return to (2.1.5). We rewrite this problem here with an additional second order term on the right-hand side.

$$dz(t)/dt = \varepsilon F(z(t), x(t), t) + \varepsilon^2 G(z(t), t), \qquad z(0) = z. \tag{2.3.1}$$

We now state the following theorem concerning the asymptotic behavior of $z(t)$.

Let F and G be smooth vector functions of their arguments and bounded together with their first and second derivatives (componentwise). Let the following limits exist uniformly in z and t_0:

$$\lim_{t\to\infty} \frac{1}{t} \int_{t_0}^{t_0+t} \int_{t_0}^{s} E\{F_i(z, x(s), s)F_j(z, x(\sigma), \sigma)\}\, d\sigma\, ds = a_{ij}(z),$$

$$\lim_{t\to\infty} \frac{1}{t} \int_{t_0}^{t_0+t} \int_{t_0}^{s} \sum_{i=1}^{n} E\left\{F_i(z, x(s), s)\frac{\partial}{\partial z_i}F_j(z, x(\sigma), \sigma)\right\} d\sigma\, ds = b_j(z), \tag{2.3.2}$$

$$\lim_{t\to\infty} \frac{1}{t} \int_{t_0}^{t_0+t} G_j(z, s)\, ds = c_j(z).$$

For fixed $z \in R^n$ and $t \geqq 0$,

$$E\{F(z, x(t), t)\} = 0. \tag{2.3.3}$$

Let U_s^t denote the σ-algebra of sets in (Ω, U, P), the probability space on which $x(t)$ is defined, with respect to which $x(\sigma)$, $s \leqq \sigma \leqq t$, is measurable. Then assume that

$$|P(AB) - P(A)P(B)| < \beta(t)P(A), \tag{2.3.4}$$

where A and B are arbitrary sets in U_0^s and U_{s+t}^∞ respectively, and

$$t^6\beta(t) \downarrow 0 \quad \text{as} \quad t \to \infty. \tag{2.3.5}$$

Let

$$\tau = \varepsilon^2 t, \qquad z^{(\varepsilon)}(\tau) = z(\tau/\varepsilon^2). \tag{2.3.6}$$

Under these hypotheses the process $z^{(\varepsilon)}(\tau)$ converges weakly to $z^{(0)}(\tau)$, $0 \leqq \tau \leqq \tau_0$, τ_0 a fixed finite constant, where $z^{(0)}$ is a diffusion Markov process with infinitesimal generator

$$\bar{V} = \sum_{i,j=1}^{n} a_{ij}(z)\frac{\partial^2}{\partial z_i \partial z_j} + \sum_{j=1}^{n} [b_j(z) + c_j(z)]\frac{\partial}{\partial z_j}. \tag{2.3.7}$$

This result is due to R. Z. Has'minskiĭ [**15**]. It was first obtained formally by R. L. Stratonovič [**5**] and it was applied extensively by him to many problems of interest in radio engineering. It was also proved, under different hypotheses, by Stratonovič in [**8**]. An operator formulation of this result was given by R. Kubo [**16**], M. Lax [**17**] and G. Papanicolaou and J. B. Keller [**18**]. A proof, under different hypotheses, by operator methods is given in [**19**].

Let us comment briefly on the hypotheses required above before proceeding with applications. The boundedness of F and G and their derivatives is actually the one requirement that rules out many problems of interest (including examples (2.2.1) and (2.2.3) of §2.2). All other hypotheses are easy to satisfy and not unrealistic. For example in (2.2.13), even if $|x(t)| < c$ almost surely, the right-hand side is not bounded as a function of $\theta \in (-\infty, \infty)$. In the version of the above theorem given in [**19**], this restriction is not imposed. Instead it is required that the Cauchy problem for the diffusion equation with $\bar{V}$ on the right-hand side have sufficiently smooth solutions. There however, $x(t)$ is a very special sort of process. We shall proceed with applications without further comments concerning the hypotheses.

Let us apply the above theorem to (2.2.3). Let H_1, H_2 be defined by

$$\begin{aligned} H_1(r, \phi, t) &= (1/\omega)h(r\cos(\omega t + \phi), -\omega r\sin(\omega t + \phi))\sin(\omega t + \phi), \\ H_2(r, \phi, t) &= (1/\omega)h(r\cos(\omega t + \phi), -\omega r\sin(\omega t + \phi))\cos(\omega t + \phi). \end{aligned} \tag{2.3.8}$$

Let $\rho(s)$ be the covariance of the stationary process $x(t)$ $(E\{x(t)\} = 0)$:

$$\rho(s) = E\{x(t + s)x(t)\}. \tag{2.3.9}$$

We assume that $|x(t)| < c$ almost surely and $\int_0^\infty s\rho(s)\,ds < \infty$ (which actually follows from (2.3.5)). Then it is easy to see that

$$\begin{aligned}\bar{V} &= \frac{1}{2\omega^2}\int_0^\infty \rho(s)\cos(\omega s)\,ds\left[\frac{\partial^2}{\partial r^2} + \frac{\partial^2}{\partial \phi^2} + \frac{\partial}{\partial r}\right]\\ &\quad + \frac{1}{2\omega^2}\int_0^\infty \rho(s)\sin(\omega s)\,ds\cdot\frac{\partial}{\partial \phi} + \bar{H}_1(r,\phi)\frac{\partial}{\partial r} + \bar{H}_2(r,\phi)\frac{\partial}{\partial \phi},\end{aligned} \tag{2.3.10}$$

$$\bar{H}_i(r,\phi) = \lim_{t\to\infty}\frac{1}{t}\int_0^\infty H_i(r,\phi,\sigma)\,d\sigma, \qquad i = 1, 2.$$

Thus, in order to obtain properties of the processes $r(\tau/\varepsilon^2)$, $\phi(\tau/\varepsilon^2)$ in the limit $\varepsilon \to 0$, $0 \leqq \tau \leqq \tau_0$, we must obtain the fundamental solution of

$$\partial W/\partial \tau = \bar{V}W, \tag{2.3.11}$$

where W is the joint transition probability density of the limit processes. Unless H_1 and H_2 turn out to be quite simple functions this is a difficult task.

Frequently [**5**] the variables r and ϕ in (2.3.11) separate and stationary (τ independent) solutions of (2.3.11) can be obtained. Since $0 \leqq \tau \leqq \tau_0$, it is not clear (mathematically) what meaning one may assign to such solutions. The question of equilibrium solutions and their meaning manifests itself also in the deterministic applications of the method of Krylov-Bogoljubov to ordinary differential equations. It is already a difficult question there [**12**, Chapter 6]. In the stochastic situation, this problem requires further investigation. Despite these difficulties however, results concerning stationary or equilibrium solutions of (2.3.11) are quite useful and physically meaningful.

Next we apply the theorem to (2.2.5). We make the same assumptions about $x(t)$ as above. Now, unless $\beta = 2\omega$ the second order terms in (2.2.5) do not contribute to $\bar{V}$. Thus we assume that $\beta = 2\omega$ and obtain

$$\begin{aligned}\bar{V} &= \frac{\omega^2}{8}\int_0^\infty \rho(s)\cos(\omega s)\,ds\,\frac{\partial^2}{\partial r^2}\\ &\quad + \left(\frac{\omega^2}{4}\int_0^\infty \rho(s)\,ds + \frac{\omega^2}{8}\int_0^\infty \rho(s)\cos(\omega s)\,ds\right)\frac{\partial^2}{\partial \phi^2}\\ &\quad + \left(\frac{\omega^2}{4}\int_0^\infty \rho(s)\cos(\omega s)\,ds + \frac{\omega\alpha}{4}\cos 2\phi\right)\frac{\partial}{\partial r}\\ &\quad + \left(\frac{\omega^2}{4}\int_0^\infty \rho(s)\sin(\omega s)\,ds + \frac{\omega\alpha}{4}\sin 2\phi\right)\frac{\partial}{\partial \phi}.\end{aligned} \tag{2.3.12}$$

When $\alpha = 0$ this is a differential operator with constant coefficients. Otherwise the diffusion equation (2.3.11) can not be solved explicitly. In [**5**] Stratonovič examines stationary solutions of (2.3.11) with $\bar{V}$ given by

(2.3.12). In view of the remarks of the preceding paragraph such solutions require further investigation.

Let us apply the theorem (2.2.13). Straightforward calculations yield

$$\begin{aligned}
\bar{V} &= \frac{k^2}{2}\int_0^\infty \rho(s)\cos(2ks)\,ds\left[\frac{\partial^2}{\partial\theta^2} + \coth\theta\,\frac{\partial}{\partial\theta} + \frac{1}{\sinh^2\theta}\,\frac{\partial^2}{\partial\psi^2}\right] \\
&\quad + \left[k^2\int_0^\infty \rho(s)\,ds + \frac{k^2}{2}\int_0^\infty \rho(s)\cos(2ks)\,ds\,\coth^2\theta\right]\frac{\partial^2}{\partial\phi^2} \\
&\quad + \left[\frac{k^2}{2}\int_0^\infty \rho(s)\,ds\,\sin(2ks)\,ds(1 - \operatorname{cosech}^2\theta)\right]\frac{\partial}{\partial\phi}\cdot \\
&\quad - \frac{k^2}{2}\left[\int_0^\infty \rho(s)\sin(2ks)\,ds\right]\frac{\partial}{\partial\phi} - \left[k^2\int_0^\infty \rho(s)\cos(2ks)\,ds\right]\frac{\coth\theta}{\sinh\theta}\,\frac{\partial^2}{\partial\phi\partial\psi}.
\end{aligned} \tag{2.3.13}$$

Note that in the variables $\psi, \theta, \bar{V}$ is simply the Laplace-Beltrami operator of the hyperbolic disc [**20**]. Since the left reflection coefficient $-e^{i\psi}\tanh\theta/2$ does not involve ϕ, it is sufficient to solve the diffusion equation in the hyperbolic disc. This was done in [**21**], where the same problem was first treated by different considerations. See also [**18**], and the work of J. A. Morrison [**22**]. As we observed in §2.2 the right reflection coefficient $e^{i\phi}\tanh\theta/2$ does not lead to a nice diffusion equation despite the fact that it decouples from the ψ variable even before the asymptotic limit. In any case, $|R|$, the modulus of the reflection coefficient, has the same behavior. The generalization of the above to coupled mode problems is carried out in [**14**] as we mentioned in §2.2.

3. **Random evolution.** Let us first convert (2.1.5) into a stochastic partial differential equation, Liouville's equation. Let

$$z(t) = g(t, s, z), \qquad t \geqq s, \tag{3.1}$$

be the solution of (2.1.5) for $t \geqq s$ and such that $z(s) = z$. Let f be a smooth function on R^n and define

$$y(t, s, z) = f[g(t, s, z)], \qquad t \geqq s. \tag{3.2}$$

It is not difficult to verify that y satisfies the partial differential equation

$$\begin{aligned}
&\frac{\partial y}{\partial s} + \varepsilon\sum_{j=1}^{n} F_j(z, x(s), s)\frac{\partial y}{\partial z_j} = 0, \qquad t > s, \\
&y(t, t, z) = f(z).
\end{aligned} \tag{3.3}$$

Let us define the following operators on $C^\infty(R^n)$, the space of infinitely differentiable functions on R^n:

$$[V(x(s), s)f](z) = -\sum_{j=1}^{n} F_j(z, x(s), s)\frac{\partial f(z)}{\partial z_j}, \qquad f \in C^\infty(R^n). \tag{3.4}$$

Using this definition we can write (3.3) in abstract form

$$dy/ds = \varepsilon V(x(s), s)y, \qquad s \leqq t, \qquad y(t) = f. \tag{3.5}$$

Let us introduce the solution operators U of the evolution equation (3.5):

$$y(s) = U(s, t)f, \qquad s \leqq t. \tag{3.6}$$

These operators are, of course, given concretely in terms of the solutions of (2.1.5) by (3.2). Note that

$$U(t_1, t_2)U(t_2, t_3) = U(t_1, t_3), \qquad t_1 \leqq t_2 \leqq t_3. \tag{3.7}$$

It is also easy to verify that

$$\frac{\partial}{\partial s}U(s, t) = \varepsilon V(x(s), s)U(s, t), \qquad s \leqq t, \tag{3.8}$$

$$\frac{\partial}{\partial t}U(s, t) = -\varepsilon U(s, t)V(x(t), t), \qquad U(t, t) = I = \text{identity}. \tag{3.9}$$

Finally we define

$$\tau = \varepsilon^2 t, \qquad \sigma = \varepsilon^2 s, \qquad U^{(\varepsilon)}(\sigma, \tau) = U(\sigma/\varepsilon^2, \tau/\varepsilon^2), \tag{3.10}$$

so that from (3.9) we have

$$\begin{gathered}\frac{\partial}{\partial \tau}U^{(\varepsilon)}(\sigma, \tau) = \frac{1}{\varepsilon}U^{(\varepsilon)}(\sigma, \tau)V(x(\tau/\varepsilon^2), \tau/\varepsilon^2), \qquad \sigma \leqq \tau, \\ U^{(\varepsilon)}(\sigma, \sigma) = I.\end{gathered} \tag{3.11}$$

We call the operators $U^{(\varepsilon)}(\sigma, \tau)$ random evolutions. This concept, quite naturally associated with stochastic equations, was introduced by Griego and Hersh [**23**] in the following manner.

Let L be a Banach space and denote its elements by f. For each $\alpha = 1, 2, \ldots, n$, let V_α denote the generator of a strongly continuous semigroup of contraction operators $T_\alpha(t)$, $t \geqq 0$, on L. Let

$$D = \bigcap_{i,j,k} \text{domain } (V_i V_j V_k)$$

be a dense set and assume that all real linear combinations of the V_α are closed operators. Let $Q = (q_{ij})$, $i, j = 1, 2, \ldots, n$, generate an ergodic Markov chain with a unique invariant probability vector

$$p_j = \lim_{t \to \infty} P_{ij}(t), \qquad (P_{ij}(t)) = e^{Qt}, \qquad i, j = 1, \ldots, n. \tag{3.12}$$

Let $t_1, t_2, \ldots, t_v$ denote the successive jump times of the sample path $x(t)$ in the interval $[\sigma/\varepsilon^2, \tau/\varepsilon^2)$, where v counts the number of jumps in this interval. Define a family of contraction operators on L by

$$U^{(\varepsilon)}(\sigma, \tau) = T_{x(\sigma/\varepsilon^2)}(\varepsilon t_1 - \sigma/\varepsilon)T_{x(t_1)}(\varepsilon t_2 - \varepsilon t_1)\cdots T_{x(t_v)}(\tau/\varepsilon - \varepsilon t_v), \qquad \sigma \leqq \tau. \tag{3.13}$$

The operators $U^{(\varepsilon)}(\sigma, \tau)$ are called random evolutions and can be considered as solutions of (3.11) with

$$V(x(s), s) = V_{x(s)}. \tag{3.14}$$

Clearly, when F in (2.1.5) is independent of t and $x(t)$ is the finite state chain of the preceding paragraph, the stochastic differential equation (2.1.5) gives rise to a random evolution (in the original sense of [**23**]) with $L = C[R^n]$, the space of bounded uniformly continuous functions on R^n, and

$$V_{x(s)} = \sum_{j=1}^{n} F_j(z, x(s))\partial/\partial z_j. \tag{3.15}$$

We need some smoothness and boundedness assumptions on F, which however can be weakened by starting with a different Banach space L. Thus random evolutions are simply operator-valued solutions of stochastic equations.

We now state the following theorem concerning the expected value $E_i\{U^{(\varepsilon)}(0, \tau)\}$ of (3.13) as $\varepsilon \to 0$, $0 < \tau \leqq \tau_0$ [**24**]. Here E_i denotes expectation with respect to the measure on the paths $x(t)$ with $x(0) = i$. Let $\bar{V}$ be defined by

$$\bar{V} = \sum_{\alpha,\beta=1}^{n} p_\alpha \int_0^\infty [P_{\alpha\beta}(u) - p_\beta]\, du\, V_\alpha V_\beta. \tag{3.16}$$

If $\bar{V}$ generates a strongly continuous semigroup on L, D and $(\beta - \bar{V})D$, $\beta > 0$, are dense in L, Q is ergodic with invariant vector $\{p_i\}$ and if

$$\sum_{\alpha=1}^{n} p_\alpha V_\alpha = 0, \tag{3.17}$$

then, for $f_i \in L$, $i = 1, \ldots, n$,

$$\lim_{\varepsilon\to 0} E_i\{U^{(\varepsilon)}(0, \tau)f_{x(\tau/\varepsilon^2)}\} = e^{\tau\bar{V}}\left(\sum_{\alpha=1}^{n} f_\alpha p_\alpha\right), \qquad 0 < \tau \leqq \tau_0. \tag{3.18}$$

Note that the limit in (3.18) is independent of i and the interval of τ is open at the left end.

It is easily seen that the theorem of §2.3, appropriately specialized, and the above result yield the same conclusions. From the point of view of

(2.1.5), (3.18) is not particularly useful since, in applications, F usually does depend explicitly on t. In [**19**], V in (3.9) is allowed to depend explicitly on t but the process $x(t)$ is even more special than a finite state chain. Further generalizations of the result (3.18), with elegant proofs, are due to T. Kurtz [**25**]. The whole subject of random evolutions is presented in detail in the work of Pinsky [**2**].

We now indicate the connection of (3.18) with the singular perturbation of partial differential equations. This demonstrates the enlarged scope of the problems associated with random evolutions. As we noted above, random evolutions were introduced in [**23**] without particular relation to stochastic equations. The interest there was the fact that

$$u_i^{(\varepsilon)}(\tau) = E_i\{U^{(\varepsilon)}(0, \tau) f_{x(\tau/\varepsilon^2)}\} \tag{3.19}$$

satisfies the following deterministic abstract Cauchy problem:

$$\frac{du_i^{(\varepsilon)}(\tau)}{d\tau} = \frac{1}{\varepsilon} V_i u_i^{(\varepsilon)} + \frac{1}{\varepsilon^2} \sum_{j=1}^{n} q_{ij} u_j^{(\varepsilon)}, \qquad u_i^{(\varepsilon)}(0) = f_i. \tag{3.20}$$

This is a consequence of a simple generalization of the Feynman-Kac formula [**26**], [**27**], [**24**]. The limit theorem (3.18) implies now that $u_i^{(\varepsilon)}(\tau)$ converges strongly to $u^{(0)}(\tau)$, $0 < \tau \leqq \tau_0$, where

$$\frac{du^{(0)}}{d\tau} = \bar{V} u^{(0)}, \qquad u^{(0)}(0) = \sum_{\alpha=1}^{n} f_\alpha p_\alpha. \tag{3.21}$$

REFERENCES

1. J. A. Morrison and J. McKenna, *Analysis of some stochastic ordinary differential equations*, SIAM-AMS Proc., vol. 6, Amer. Math. Soc., Providence, R.I., 1973, pp. 97–162.

2. M. Pinsky, *Multiplicative operator functionals and their asymptotic properties*, Advances in Probability (to appear).

3. G. E. Uhlenbeck and L. S. Ornstein, *On the theory of the Brownian motion*, Phys. Rev. **36** (1930), 823–841.

4. M. C. Wang and G. E. Uhlenbeck, *On the theory of the Brownian motion*. II, Rev. Modern Phys. **17** (1945), 323–342. MR **7**, 130.

5. R. L. Stratonovič, *Topics in the theory of random noise*. Vols. 1, 2, Izdat. "Sovetskoe Radio," Moscow, 1961; English transl., Gordon and Breach, New York, 1963. MR **28** #1660.

6. U. Frisch, *Wave propagation in random media*, Probabilistic Methods in Appl. Math., vol. 1, Academic Press, New York, 1968, pp. 75–198. MR **42** #4088.

7. H. P. McKean, Jr., *Stochastic integrals*, Probability and Math. Statist., no. 5, Academic Press, New York, 1969. MR **40** #947.

8. R. L. Stratonovič, *Conditional Markov processes and their application to the theory of optimal control*, Izdat. Moskov. Univ., Moscow, 1966; English transl., American Elsevier, New York, 1967. MR **33** #5391.

9. R. Z. Has'minskiĭ, *Stability of systems of differential equations under random perturbations of their parameters*, "Nauka," Moscow, 1969; English transl., Transl. Math. Monographs, Amer. Math. Soc., Providence, R.I. (to appear). MR **41** #3925.

10. H. Bunke, *Über Stabilitätseigenschaften im Mittel von Differentialgleichungs-systemen mit stochastischen Parametern*, Monatsh. Deutsch. Akad. Wiss. Berlin **12** (1970), 734–740. MR **43** #2316.

11. N. Krylov and N. N. Bogoljubov, *Introduction to non-linear mechanics*, Izdat. Akad. Nauk SSSR, Moscow, 1937; English transl., Ann. of Math. Studies, no. 11, Princeton Univ. Press, Princeton, N.J., 1943. MR **4**, 142.

12. N. N. Bogoljubov and Ju. A. Mitropol'skiĭ, *Asymptotic methods in the theory of non-linear oscillations*, Fizmatgiz, Moscow, 1958; English transl., Gordon and Breach, New York, 1961. MR **20** #6812; MR **25** #5242.

13. Ju. A. Rozanov, *Stationary random processes*, Fizmatgiz, Moscow, 1963; English transl., Holden-Day, San Francisco, Calif., 1967. MR **28** #2580; MR **35** #4985.

14. R. Burridge and G. C. Papanicolaou, *The geometry of coupled mode propagation in one dimensional random media*, Comm. Pure Appl. Math., **25** (1972) (to appear).

15. R. Z. Has'minskiĭ, *A limit theorem for solutions of differential equations with a random right hand part*, Teor. Verojatnost. i Primenen. **11** (1966), 444–462 = Theor. Probability Appl. **11** (1966), 390–406. MR **34** #3637.

16. R. Kubo, *Stochastic Liouville equation*, J. Mathematical Phys. **4** (1963), 174–183. MR **26** #7370.

17. M. Lax, *Classical noise*. IV. *Langevin methods*, Rev. Modern Phys. **38** (1966), 561–566.

18. G. C. Papanicolaou and J. B. Keller, *Stochastic differential equations with applications to random harmonic oscillators and wave propagation in random media*, SIAM J. Appl. Math. **21** (1971), 287–305.

19. G. C. Papanicolaou and R. Hersh, *Some limit theorems for stochastic equations and applications*, Indiana Univ. Math. J. **21** (1972), 815–840.

20. C. Carathéodory, *Conformal representation*, Cambridge Univ. Press, New York, 1958.

21. M. E. Gercenšteĭn and V. B. Vasil'ev, *Waveguide with random nonhomogeneities and Brownian motion on the Lobachevsky plane*, Teor. Verojatnost. i Primenen. **4** (1959), 424–432 = Theor. Probability Appl. **4** (1959), 392–398. MR **22** #9071.

22. J. A. Morrison, *Application of a limit theorem to solutions of a stochastic differential equation*, J. Math. Anal. Appl. **39** (1972), 13–36.

23. R. Griego and R. Hersh, *Theory of random evolutions with applications to partial differential equations*, Trans. Amer. Math. Soc. **156** (1971), 405–418. MR **43** #1261.

24. R. Hersh and G. C. Papanicolaou, *Non-commuting random evolutions and an operator valued Feynman-Kac formula*, Comm. Pure Appl. Math. **25** (1972), 337–367.

25. T. G. Kurtz, *A limit theorem for perturbed operator semigroups with applications to random evolutions* (to appear).

26. M. Kac, *On the distributions of certain Wiener functionals*, Trans. Amer. Math. Soc. **65** (1949), 1–13. MR **10**, 383.

27. Ju. L. Daleckiĭ, *Functional integrals associated with operator evolution equations*, Uspehi Mat. Nauk **17** (1962), no. 5 (107), 3–115 = Russian Math. Surveys **17** (1962), no. 5, 1–107. MR **28** #4389.

28. G. C. Papanicolaou, *Wave propagation in a one-dimensional random medium*, SIAM J. Appl. Math. **21** (1971), 3–8.

COURANT INSTITUTE OF MATHEMATICAL SCIENCES, NEW YORK UNIVERSITY

Wave Propagation and Conductivity in Random Media

Melvin Lax

Abstract. The variety of mathematical methods used to derive the density of eigenstates and the conductivity of a random medium are summarized and compared. Previously unpublished material on the exact coherent potential, the quasicrystalline potential with modified propagators, and the pair approximation to a.c. impurity band conduction is included.

1. **Introduction.** It is appropriate that a physicist be chosen to discuss wave propagation and conductivity in random media at this symposium. Random media of several kinds—alloys, liquid metals, impurity bands in semiconductors and amorphous semiconductors—have recently assumed increasing importance in physics. Moreover, there has been a surge of activity on these topics in the past few years. Because of the difficulty of the problems, a wide variety of mathematical methods has been employed: node counting by Monte Carlo methods, Markoff methods, multiple scattering methods including what is known as the coherent potential approximation, counting of potential fluctuation minima, path integral methods, the use of renormalized perturbation series to distinguish between extended and localized wave functions, percolation theory and continuous time random walk methods. These techniques will be reviewed briefly in this paper. The node counting and Markoff methods are exact but restricted to one dimension. The random walk method of Scher and Lax [**1972**] makes an *exact* solution of a model three-dimensional problem. This is the usual ploy of a mathematician. Theoretical physicists are more inclined to make an *approximate* solution of the original problem. The problems treated here are sufficiently difficult that, in most cases, even the model problems are treated approximately.

Physicists are concerned with the "density of states", i.e., the distribution

---

AMS (*MOS*) *subject classifications* (1970). Primary 60H10, 60H15, 60J15, 60J25.

of eigenvalues of a stochastic differential equation, and with the electrical conductivity which is expressible in terms of the average of the product of two Green's functions. These are more difficult to calculate than the mean wave function or low order moments of the wave function or its derivatives. Moreover, we are concerned with three-dimensional problems. Thus the exact solutions of one-dimensional examples discussed by Papanicolaou and by Morrison and McKenna in these Proceedings are of interest to us only as testing grounds for the approximate methods to be used in three dimensions. These articles should be referred to for work and references almost orthogonal to ours.

The Markoffian methods used to solve one-dimensional problems are not generalizable to three dimensions. But they can yield exact solutions for later comparison. We shall therefore briefly review in §2 the work of Lax and Phillips [**1958**], who formulate a suitable one-dimensional problem and provide a Monte Carlo solution of it. The exact results for the same problem, obtained by Frisch and Lloyd [**1960**] using Markoff techniques, are also presented. In later sections we shall illustrate a variety of other mathematical techniques found useful, without attempting a detailed survey of the literature.

2. **One-dimensional impurity band.**

A. *Formulation.* An impurity band is a set of closely spaced energy eigenvalues of an electron in a solid due to the presence of impurities. Each impurity atom is assumed to have a potential strong enough to produce a bound or localized state of the electron. All the essential ingredients of the impurity band problem are contained in the Lax-Phillips [**1958**] model of a set of Dirac delta function potentials distributed independently and uniformly with mean density n on a line. The Schrodinger equation for such a problem can be reduced to

$$\left[\frac{d^2}{dx^2} + 2K_0 \sum_j \delta(x - x_j)\right]\psi(x) = K^2\psi(x). \tag{2.1}$$

The problem is to find the number $\nu(n, K)$ of bound states, i.e., eigenfunctions of (2.1), with eigenvalues $\leq K$, as a function of n and K. For a single impurity at the origin, the localized eigenstate is of the form

$$\psi(x) \sim \exp(-K_0|x|) \tag{2.2}$$

so that the radius of this localized state is $1/K_0$. The dimensionless parameter of the problem is

$$\varepsilon = n/K_0, \tag{2.3}$$

the average number of impurities to be found within the radius of a single impurity. We shall henceforth use $(K_0)^{-1}$ as a unit of length. In the original units the energy is given by

$$E = -\hbar^2 K^2/(2m) \tag{2.4}$$

where $\hbar$ is Planck's constant $h/2\pi$ and m is the electron mass so that K has the same dimensions as K_0, an inverse length.

B. *Monte Carlo* (*node counting*) *methods.* The distribution of energy eigenvalues K_m can be obtained with the help of the Sturm comparison theorem. This theorem implies that the number of eigenstates of (2.1) with $K_m < K$ is determined (within ± 1) by the number of zeroes of $\psi(x, K)$. To use this fact we specify K and $\psi'(0)/\psi(0)$ and then solve (2.1) for $\psi(x, K)$ with a particular choice of $x_1, \ldots, x_{\mathcal{N}}$. The point x_1 is chosen at random using the nearest neighbor distribution $n \exp(-nx)$, and the intervals $x_2 - x_1, x_3 - x_2, \ldots, x_{\mathcal{N}} - x_{\mathcal{N}-1}$ are chosen in the same way. The total number of zeroes of $\psi(x, K)$ between $x = 0$ and $x = x_{\mathcal{N}}$ is expected to grow linearly with $\mathcal{N}$. Therefore we calculate

$$N(K) = \text{number of zeroes}/\mathcal{N}. \tag{2.5}$$

Since the process of locating the points x_j is ergodic, $N(K)$ approaches a limit as $\mathcal{N}$ increases so there is no need to take an ensemble average.

The cumulative density of states $N(K)$ was determined by this method for a number of values of ε from $\varepsilon = 0.01$ to $\varepsilon = 10$. Results for $\varepsilon = 0.1$ and $\varepsilon = 10$ are given in Tables I and II. Examination of Table I shows that 14.2% of the states lie between $K = .999995$ and 1.000005. For $\varepsilon = 0.01$, 22.9% of the states lie between $1 - 10^{-40}$ and $1 + 2 \times 10^{-49}$. It is difficult to plot these results except against the unusual variable

$$x = |K - 1|^{n/K}, \tag{2.6}$$

as shown in Figure 1.

C. *Pair approximation.* The unusual nature of these results calls for some comment. At very low densities, $\varepsilon \ll 1$, we might expect the density of states (per atom) to approximate that of an isolated atom which is $\delta(K - 1)$. The presence of nearest neighbor atoms produces a perturbation which spreads this delta function into a function with an algebraic singularity of the form $|K - 1|^{-1+2\varepsilon}$. Lax and Phillips treated a pair of neighboring atoms exactly (using multiple scattering techniques) and then averaged over the distribution of nearest neighbor distances to obtain the result

$$\begin{aligned} N(K) &= \tfrac{1}{2} + \tfrac{1}{2}x^2, \qquad 0 \leqq K \leqq 1, \\ &= \tfrac{1}{2} - \tfrac{1}{2}x^2, \qquad 1 \leqq K \leqq 2. \end{aligned} \tag{2.7}$$

where x is given by (2.6).

TABLE I
Integrated density of states of negative energies, $\varepsilon = 0.1$

κ/κ_0	Machine results (%) (500 atoms)	Machine results (%) (1000 atoms)	Pair theory (%)	Schmidt's formula (%)
0.001	89.6	89.2	91.0	90.8
0.67	84.6	84.6	85.7	86.4
0.92	78.6	77.9	78.8	79.4
0.99	70.6	69.7	70.0	71.0
0.997	65.6		65.6	66.9
0.999	62.6		62.6	64.1
0.9995	61.4	61.1	60.9	62.3
0.99968	60.2	59.9	60.0	61.3
0.99985	58.6	58.6	58.5	59.9
0.999995	53.8	53.2	54.4	54.7
0.999999	51.4	51.4	53.0	52.9
1.000000	42.8		50.0	44.4
1.000005	36.6	37.0	45.6	38.5
1.00032	32.8	33.4	40.0	33.8
1.0010	31.2	31.8	37.4	32.0
1.0030	29.0		34.6	28.6
1.01	25.4	25.9	30.0	26.0
1.078	18.6	18.2	18.9	17.2
1.33	6.4	7.7	7.9	7.3
2.00	0.4	0.5	0.0	0.0

Examination of Figure 1 shows that the pair approximation is excellent for K away from 1. Indeed, for any small but finite $K - 1$, the pair approximation becomes asymptotically correct as ε approaches zero. Any discrepancies lie in a region about $K = 1$ of order $\exp(-1/\varepsilon)$. In this region, however, sizable errors can occur. Indeed, in the pair approximation $N(1) = \frac{1}{2}$ whereas the exact result for infinitesimal ε is $N(1) = \frac{4}{9}$, as is shown in Figure 1 and Table I. The reason for this discrepancy is that the very center of the line $K = 1$ is produced by pairs that are infinitely separated, so that three particle and four particle contributions, etc., do not fall off as ε^3 and ε^4 but simply become small numerically. A detailed discussion of this point has been given by Lifshitz [**1965**].

D. *Integral equation method.* A more accurate treatment in the immediate vicinity of $K = 1$ was obtained by Schmidt [**1957**] using an integral

TABLE II

Integrated density of states at negative energies, $\varepsilon = 10$

κ/κ_0	Machine results (%)	Optical model (%)	Local density model (%)
0.00001	14.1	14.1	14.1
1.00	13.8	13.8	13.8
2.00	12.4	12.6	12.4
2.83	10.9	10.9	10.5
3.50	9.6	8.8	8.1
4.16	6.7	5.2	4.9
4.47	4.85	0.0	3.4
4.97	3.2	0.0	1.3
5.50	1.5	0.0	0.3
6.00	0.7	0.0	0.0

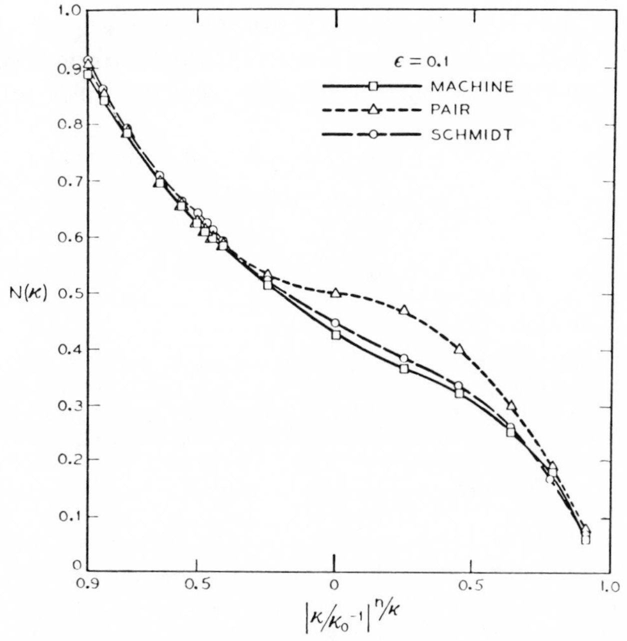

FIGURE 1

Integrated density of states in a one-dimensional impurity band at the low concentration $\varepsilon = n/K_0 = 0.1$ plotted against the unusual variable $x = |y - 1|^{\varepsilon/y}$ where $y = K/K_0$ = [energy/binding energy of isolated impurity]$^{1/2}$. (Taken from Lax-Phillips [**1958**].)

equation relating the distribution of values of ψ'/ψ at one delta function to its value at the next. A more careful solution of this integral equation by Lax and Phillips [**1958**], uniformly valid in K for small ε, led to the result

$$(2.8)\qquad \begin{aligned} N(K) &= (\tfrac{3}{2} - \tfrac{1}{2}x)^{-2}, && 0 \leqq K \leqq 1, \\ &= (1 - x)(\tfrac{3}{2} - \tfrac{1}{2}x)^{-2}, && 1 \leqq K \leqq 2. \end{aligned}$$

The original Schmidt formula has the form (2.8) with $x = |K - 1|^{\varepsilon}$. That these formulae are appreciably different can be seen by evaluating them at $K = 0$. Under Schmidt's choice, $x = 1$,

$$(2.9)\qquad N_{\text{Schmidt}}(0) = 1.$$

Under our modified Schmidt formula, at $K = 0$, equation (2.6) yields $x = \exp(-\varepsilon)$ and

$$(2.10)\qquad N_{\text{MS}}(0) = 1 - \varepsilon + \tfrac{5}{4}\varepsilon^2 - \cdots,$$

where $N_{\text{MS}}(K)$ is the modified Schmidt formula (2.8) combined with (2.6). It can be seen from Table I or Figure 1 that the pair formula (2.7) is in good agreement with the exact results except in the immediate vicinity of the center, $K = 1$, of the density of states, but the modified Schmidt formula (2.8) is in excellent agreement over the whole range. Morrison [**1962**] has supplied a rigorous proof that our modified Schmidt formula is indeed the correct low density limit.

The differential or ordinary density of states can be defined by

$$(2.11)\qquad n(K) = \frac{dN(K)}{dK} = \frac{dN(x)}{dx}\frac{dx}{dK}.$$

Since

$$(2.12)\qquad \frac{dx}{dK} = \frac{n}{K}\frac{1}{|K - 1|^{1-n/K}}\left[1 - \left(1 - \frac{1}{K}\right)\ln|1 - K|\right]$$

is singular at $K = 1$ and dN/dx from equation (2.8) is not, the differential density in the immediate neighborhood of $K = 1$ is given by

$$(2.13)\qquad n(K) = C(n/K)|K - 1|^{n/K-1},$$

where C has different values on the two sides of the singularity. The width of the singular region which contains a fixed percentage of the states is thus of order $\exp(-1/\varepsilon)$ which is quite narrow for $\varepsilon = 0.1$. For larger ε, say $\varepsilon \geqq 1$, no sign of such a singularity appears to be present (see Figure 2).

E. *Markoff methods.* For the case of delta function potentials, Frisch and Lloyd [**1960**] have shown that ψ and $d\psi/dx$ behave as a pair of Markoffian variables describable by a distribution function, $P(\psi, d\psi/dx, x)$,

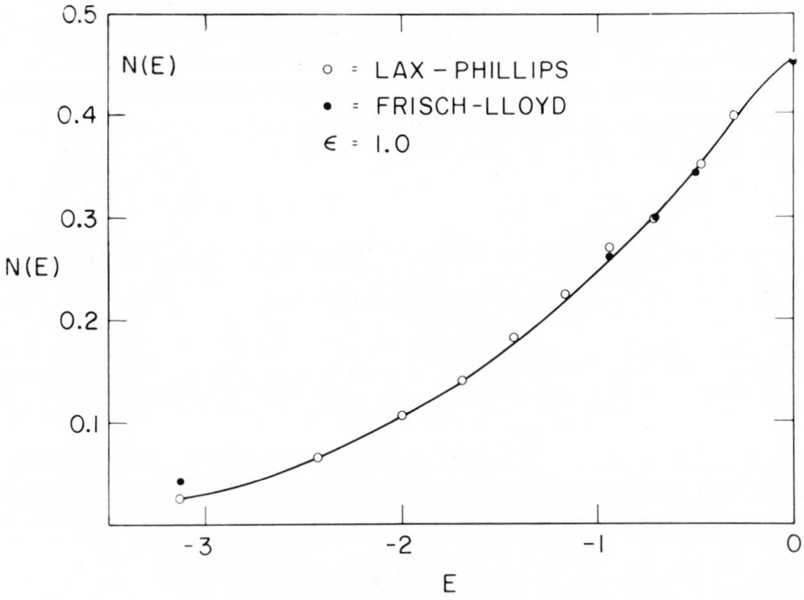

FIGURE 2

Integrated density of states in a one-dimensional impurity band at the medium concentration $\varepsilon = n/K_0 = 1.0$ plotted against the dimensionless energy $E = -\frac{1}{2}(K/K_0)^2 = \frac{1}{2}$ energy/binding energy of isolated impurity. No singularity is visible at the energy $E = -\frac{1}{2}$ of isolated impurities. Circles are Monte Carlo results of Lax and Phillips [**1958**], and black dots are exact Markoff results of Frisch and Lloyd [**1960**].

where x plays the role of time. The stationary (independent of x) solution for P can be used to count the expected number of nodes on a given line interval and thus to obtain the cumulative density of states. Since the Markoff method is the primary method of attack discussed in the review papers in this volume by Papanicolaou, and by Morrison and McKenna, we shall rely on these papers and on the reviews by Keller [**1962**], [**1964**] to provide a summary of the mathematical literature. We mention however that Halperin [**1965**] has an application of the Frisch-Lloyd method to the Gaussian case, and Lax [**1966**, §4] has a succinct summary of the Frisch-Lloyd-Halperin work.

Recent work of Morrison [**1972**], McKenna and Morrison [**1971**], and Papanicolaou and Keller [**1971**] are based on a mathematical theorem of Has'minskiĭ [**1966**] that is equivalent to the reduction of a short correlation problem to a Markoffian problem made by Lax [**1966**, §5].

3. **Multiple scattering approach.** The multiple scattering approach to be discussed now has the advantage over some other methods that in it the interaction between the wave and each single scatterer is treated exactly.

A. *Single scattering.* We shall treat first single scattering using abstract Hilbert vectors, as well as the customary space representation approach, to prepare for the abstract treatment of multiple scattering to be given later. The abstract and concrete Schrodinger equations can be written in the respective forms

$$(E - H_0 - v)\Psi = 0, \qquad [k^2 + \nabla^2 - v(\boldsymbol{r})]\psi(\boldsymbol{r}) = 0. \tag{3.1}$$

These equations can be converted to the corresponding integral forms

$$\Psi_a = \Phi_a + (E - H_0)^{-1} v\Psi_a \tag{3.2}$$

or

$$\psi_a(r) = \exp(i\boldsymbol{a} \cdot \boldsymbol{r}) - \frac{1}{4\pi} \int \frac{e^{ik|\boldsymbol{r} - \boldsymbol{r}'|}}{|\boldsymbol{r} - \boldsymbol{r}'|} v(\boldsymbol{r}')\, \psi_{\boldsymbol{a}}(\boldsymbol{r}')\, d\boldsymbol{r}', \tag{3.3}$$

where the total wave is labelled by the k-vector $\boldsymbol{a}$ of the plane incident wave with which it is associated. The second term in (3.3), the scattered wave, reduces at large distances to

$$\psi_{\text{scatt}}(\boldsymbol{r}) \to -\exp(ikr)t_{\boldsymbol{ba}}/4\pi r \tag{3.4}$$

where the matrix element $t_{\boldsymbol{ba}}$ of the transition operator t is given by

$$t_{\boldsymbol{ba}} = (\varphi_{\boldsymbol{b}}(\boldsymbol{r}'), v(\boldsymbol{r}')\, \psi_{\boldsymbol{a}}(\boldsymbol{r}')), \tag{3.5}$$

$$\varphi_{\boldsymbol{b}}(\boldsymbol{r}') = \exp(i\boldsymbol{b} \cdot \boldsymbol{r}') \quad \text{and} \quad \boldsymbol{b} = k\boldsymbol{r}/r \tag{3.6}$$

is a wave vector in the direction of observation. In abstract terms, the transition operator can be defined by

$$v\Psi_{\boldsymbol{a}} = t\Phi_{\boldsymbol{a}}, \tag{3.7}$$

and in view of (3.2), it obeys the integral equation

$$t = v + v(E - H_0)^{-1} t. \tag{3.8}$$

In all cases, we understand E to be $E + i\eta$ (where η is infinitesimal) in order that $(E - H_0)^{-1}$ represent the outgoing Green's function. In our abstract notation, the scattered wave [the second term in (3.2)] can be written using (3.7) in the form

$$\begin{aligned} \Psi_{\text{scatt}} &= (E - H_0)^{-1} t\Phi_{\boldsymbol{a}} \\ &= (E - H_0)^{-1} t \quad \text{(incident wave).} \end{aligned} \tag{3.9}$$

Our calculation, so far, has been made for a scatterer located at the origin. For a scatterer located at $\boldsymbol{j}$ we replace t by t_j whose matrix elements are given by

$$(t_j)_{\boldsymbol{ba}} = \exp[i(\boldsymbol{a} - \boldsymbol{b}) \cdot \boldsymbol{j}] t_{\boldsymbol{ba}} \tag{3.10}$$

(see Lax [**1951**]). This result can be understood by examining Figure 3 in which we see that $\boldsymbol{a} \cdot \boldsymbol{j}$ is the extra phase required for the incident wave to reach $\boldsymbol{j}$ and $\boldsymbol{b} \cdot \boldsymbol{j}$ is the reduction in phase for the scattered wave to get from $\boldsymbol{j}$ to the observation point (both compared to a scatterer at the origin).

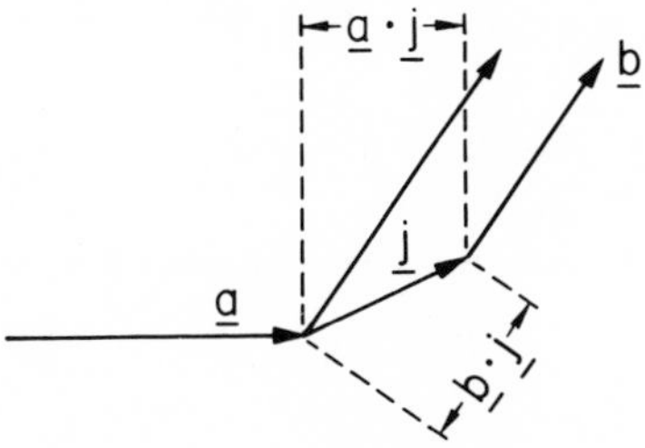

FIGURE 3

A scatterer at $\boldsymbol{j}$ has the extra phase $\boldsymbol{a} \cdot \boldsymbol{j}$ compared to a scatterer at the origin, and a reduced phase lag $\boldsymbol{b} \cdot \boldsymbol{j}$ in the scattered wave for a net phase lag of $\exp[i(\boldsymbol{a} - \boldsymbol{b}) \cdot \boldsymbol{j}]$.

B. *Multiple scattering*: *Forward scattering*. The general three-dimensional multiple scattering problem we wish to solve can be written in the abstract form

$$\left[E - H_0 - \sum_j v_j\right]\Psi = 0, \qquad v_j = v(\boldsymbol{r} - \boldsymbol{j}). \tag{3.11}$$

In view of the relation (3.9) between scattered wave and incident wave, Lax [**1951**] wrote down, intuitively, the following set of multiple scattering equations

$$\Psi = \Phi + \sum_j (E - H_0)^{-1} t_j \Psi^j, \tag{3.12}$$

$$\Psi^j = \Phi + \sum_{i \neq j} (E - H_0)^{-1} t_i \Psi^i. \tag{3.13}$$

Here Ψ is to be interpreted as the total wave and Ψ^j the wave which excites scatterer j, so that

$$\text{wave leaving } j = (E - H_0)^{-1} t_j \ \text{(wave exciting } j) \tag{3.14}$$

in accord with (3.9). Moreover, the wave Ψ^j which excites j includes the external driving wave Φ, if any, plus the waves which leave all other scatterers i.

Since the equations (3.12) and (3.13) have been rederived with the help of infinite expansions (see, for example, Watson [**1953**]) we shall include in Appendix A a simple closed form derivation of these equations.

If we take the ensemble average of equation (3.12), it can be rewritten in a form containing a coherent potential V_c:

$$(E - H_0 - V_c)\langle\Psi\rangle = 0 \tag{3.15}$$

providing we define

$$V_c = \sum_j \langle t_j \Psi^j\rangle / \langle\Psi\rangle. \tag{3.16}$$

The coherent potential V_c of solid state physics is also the optical potential of nuclear physics (Feshbach [**1958**]).

If we make the effective field approximation (EFA)

$$\Psi^j \approx \langle\Psi\rangle, \tag{3.17}$$

equation (3.16) leads to

$$V_c \approx \sum_j \langle t_j\rangle \tag{3.18}$$

or in matrix form

$$(V_c)_{\boldsymbol{ba}} = \int n(\boldsymbol{j})\, d\boldsymbol{j}\, e^{i(\boldsymbol{a}-\boldsymbol{b})\cdot \boldsymbol{j}} t_{\boldsymbol{ba}} = n\delta_{\boldsymbol{ba}} t_{\boldsymbol{ba}}, \tag{3.19}$$

where $n(\boldsymbol{j}) = n$ is the uniform density of scatterers. This result (3.19) that the coherent potential is diagonal and (to a good approximation) equal to the forward scattering transition matrix is the principal result of Lax [**1951**].

The approximation (3.17) was first made by Foldy [**1945**] in connection with a treatment of isotropic scatterers. A way of eliminating the approximation (3.17) and taking account of the ratio between the effective exciting field and the average field, which includes the effect of correlation between scatterer positions, was made by Lax [**1952**]. This improved approximation, referred to as the quasicrystalline approximation, is discussed in part E of this section and Appendix B.

C. *Exact state density and exact* V_c. The total density of states can be defined by

$$n(E) = \left\langle \sum_j \delta(E - E_j)\right\rangle \tag{3.20}$$

since

$$N(E'') - N(E') = \int_{E'}^{E''} n(E)\, dE = \left\langle \sum_j \begin{matrix} 1 & \text{if } E' < E_j < E'' \\ 0 & \text{otherwise} \end{matrix} \right\rangle \tag{3.21}$$

yields the correct number of states between the E', E'' limits. Since equation (3.20) is a trace, an invariant expression for the density of states is

$$n(E) = \mathrm{tr}\left\langle \delta\left(E - H_0 - \sum_j v_j \right) \right\rangle. \tag{3.22}$$

To take advantage of the coherent potential V_c it seems logical to rearrange the perturbation by writing

$$H = H_0 + V_c, \qquad V_j = v_j - V_c/N, \qquad V = \sum_j V_j \tag{3.23}$$

so that

$$n(E) = -(1/\pi)\,\mathrm{Im}\,\mathrm{tr}\langle G \rangle, \tag{3.24}$$

where

$$G = (E - H - V)^{-1}, \tag{3.25}$$

where we understand E to have an infinitesimal positive imaginary part. (This choice results in G being an outgoing Green's function as in (3.3).) The Green's function obeys the equation

$$(E - H - \sum V_j)G = 1. \tag{3.26}$$

By the intuitive procedure of this section, or the algebraic procedure of Appendix A, equation (3.26) is converted to the multiple scattering form:

$$G = G_c + \sum_j (E - H)^{-1} T_j G^j, \tag{3.27}$$

$$G^j = G_c + \sum_{i \neq j} (E - H)^{-1} T_i G^i, \tag{3.28}$$

where

$$G_c = (E - H)^{-1} = (E - H_0 - V_c)^{-1} \tag{3.29}$$

is the Green's function describing propagation in the medium with the coherent potential V_c and the transition operator obeys

$$T_j = V_j + V_j (E - H)^{-1} T_j. \tag{3.30}$$

The ensemble average of equation (3.27) leads to

$$\langle G \rangle = G_c = (E - H_0 - V_c)^{-1} \tag{3.31}$$

providing V_c is chosen to obey

$$\sum_j \langle T_j G^j \rangle = 0. \tag{3.32}$$

Since no approximation has been made thus far, the coherent potential has been defined *exactly* by equations (3.31) and (3.32).

The density of states for the case of a homogeneous distribution of scatterers, in which case V_c is diagonal in the k representation, can be obtained by combining equations (3.24) and (3.31):

$$n(E) = -\frac{L^d}{\pi} \operatorname{Im} \int \frac{d\mathbf{k}}{(2\pi)^d} \frac{1}{E + i\eta - \frac{1}{2}k^2 - V^c_{\mathbf{kk}}}, \tag{3.33}$$

where we have used units in which $\hbar = m = K_0 = 1$, $V^c_{\mathbf{kk}}$ is the diagonal element of the coherent potential V_c, $d = 1, 2$, or 3 is the dimension of the space and L^d is the volume of the crystal.

To implement equation (3.32) we need an evaluation of T_j. The exact but formal algebraic solution of equation (3.30) is

$$T_j = V_j + V_j(E - H - V_j)^{-1}V_j, \tag{3.34}$$

which can be written more explicitly as

$$\begin{aligned} T_j &= v_j - (V_c/N) \\ &\quad + \left(v_j - \frac{V_c}{N}\right)\left(E - H_0 - \frac{N-1}{N}V_c - v_j\right)^{-1}\left(v_j - \frac{V_c}{N}\right). \end{aligned} \tag{3.35}$$

In the "thermodynamic" limit in which N approaches infinity,

$$T_j = \hat{t}_j - V_c/N \tag{3.36}$$

where

$$\hat{t}_j = v_j + v_j(E - H_0 - V_c - v_j)^{-1}v_j. \tag{3.37}$$

One term of order $1/N$ has been retained in equation (3.35) because only this term when summed over j gives a contribution of order unity.

By inserting (3.36) into (3.33), we obtain the exact "thermodynamic" result

$$V_c = N \sum_j \langle \hat{t}_j G^j \rangle \Big/ \sum_j \langle G^j \rangle, \tag{3.38}$$

where $\hat{t}$ is a solution of the integral equation

$$\hat{t}_j = v_j + v_j e^{-1} \hat{t}_j \tag{3.39}$$

containing the modified propagator

$$e = E - H_0 - V_c. \tag{3.40}$$

In a liquid or an impurity band, homogeneity guarantees as in (3.19) that the numerator in (3.38), the denominator and V_c are all diagonal in $\mathbf{k}$.

D. *The coherent potential approximation.* If we neglect the difference between the exciting field and the average field and make the crude approximation

(3.41) $$G^j \approx G_c,$$

the rigorous condition (3.32) reduces to the simpler condition

(3.42) $$\sum_j \langle T_j \rangle = 0$$

which we shall refer to as the coherent potential approximation (CPA).

In view of equation (3.36), equation (3.42) can be reduced to a self-consistent approximation for the coherent potential

(3.43) $$V_c = V_{\text{CPA}} = \sum_j \langle \hat{t}_j \rangle,$$

whose matrix elements as in (3.19) are given by

(3.44) $$(V_c)_{\boldsymbol{ba}} = n\delta_{\boldsymbol{ba}}\hat{t}_{\boldsymbol{ba}}.$$

This is identical to the modified propagator formalism of Watson [**1953**]. It is derived, however, without using infinite expansions. Moreover, the exact condition (3.38) can be used to take account of correlations (see Appendix B) or to find improved approximations.

It was knowledge of the modified propagator results (3.39) and (3.44) that led the author to suggest to J. R. Klauder that he investigate the accuracy of this approximation by comparison with the Monte Carlo results of Lax and Phillips [**1958**] or the exact Markoff results of Frisch and Lloyd [**1960**].

Instead Klauder [**1961**] made an elegant field theoretic reformulation of the problem. He then used the diagram summation techniques of field theory to obtain six approximate solutions for the density of states. He gives no numerical results for the sixth solution. His second and third solutions are more closely related to perturbation procedures and will not be discussed here. The relation between his first, fourth and fifth solutions and our multiple scattering procedure will be clarified below.

With $n(E)$ now representing the density of states on a per atom basis, equation (3.33) can now be rewritten in one dimension as

(3.45) $$n(E) = -\frac{1}{\pi\varepsilon}\operatorname{Im}\int\frac{dk}{2\pi}\frac{1}{E^+ - V^c_{\boldsymbol{kk}} - \frac{1}{2}k^2},$$

where $E^+ = E + i0$. The simplest approximation to V^c is the virtual crystal or average potential. If $V^c_{\boldsymbol{kk}}$ is a function only of E, as will be seen below, for delta function potentials

(3.46) $$n(E) = \frac{1}{\pi\varepsilon}\operatorname{Re}\frac{1}{[2(E - V_c)]^{1/2}}.$$

The simplest approximation to V_c is the virtual crystal (or average) potential model, erroneously called the optical model by Lax and Phillips

[**1958**]. It yields

$$V_{VC} = \left\langle \sum_j v(x - x_j) \right\rangle = n \int v(x)\, dx = -\varepsilon, \tag{3.47}$$

where $\varepsilon = n/K_0$ is the dimensionless density and we have used

$$H_0 = -\tfrac{1}{2}d^2/dx^2, \qquad v(x) = -K_0\delta(x) = -\delta(x) \tag{3.48}$$

and the units $\hbar = m = K_0 = 1$. This virtual crystal model yields

$$n_{VC}(E) = \frac{1}{\pi\varepsilon} \frac{1}{[2(E + \varepsilon)]^{1/2}}, \tag{3.49}$$

which is Klauder's first approximation.

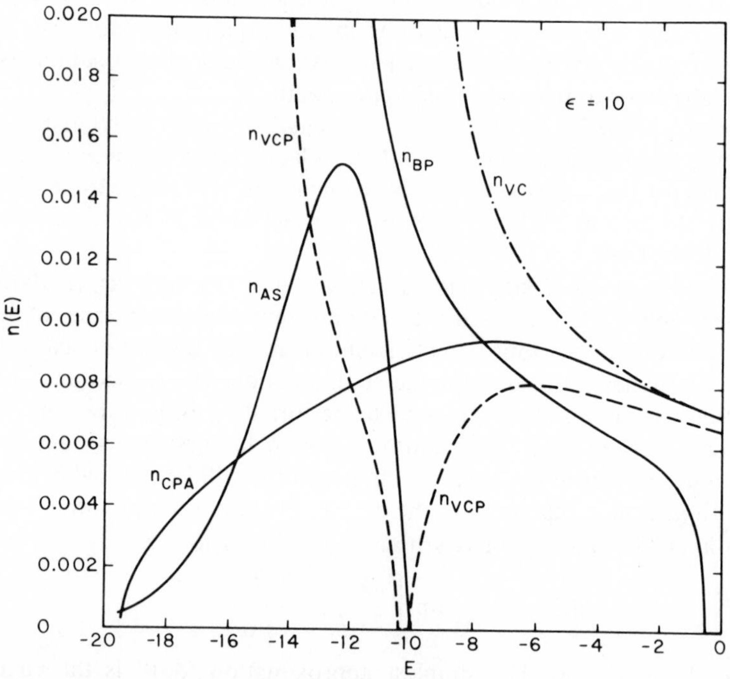

FIGURE 4

Various approximations to the differential density of states $n(E)$ in a one-dimensional impurity band at high concentration $\varepsilon = n/K_0 = 10$. The curves are labelled VC = virtual crystal approximation, BP = bare propagator = average t matrix approximation (ATA), VCP = virtual crystal propagator approximation = average t matrix approximation when mean potential is absorbed into unperturbed Hamiltonian, CPA = coherent potential approximation = average t matrix with selfconsistently determined propagator, AS = asymptotic form (3.59) of a solution to a Gaussian delta correlated model of §4.

Better approximations are obtained by expressing V_c in terms of the forward scattering t matrix:

$$V_c = \varepsilon \hat{t}_{aa} \tag{3.50}$$

where the solution of equation (3.39) (see Appendix C) is given by

$$\hat{t} = -\delta(x)(1 - [-2(E - V_c)]^{-1/2})^{-1} \tag{3.51}$$

so that

$$V_c = \varepsilon \hat{t}_{aa} = -\varepsilon(1 - [-2(E - V_c)]^{-1/2})^{-1}. \tag{3.52}$$

Different levels of approximation can now be obtained with different choices of V_c on the right-hand side of (3.52). The bare propagator sets $V_c = 0$ and obtains

$$V^c_{\mathrm{BP}} = -\varepsilon[1 - (-2E)^{-1/2}]^{-1}. \tag{3.53}$$

The use of the virtual crystal potential $V_c = -\varepsilon$ on the right-hand side yields the virtual crystal propagator

$$\begin{aligned} V^c_{\mathrm{VCP}} &= -\varepsilon(1 - [-2(E + \varepsilon)]^{-1/2})^{-1} \\ &= -\varepsilon(1 - i[2(E + \varepsilon)]^{-1/2})^{-1}. \end{aligned} \tag{3.54}$$

This is Klauder's fourth approximation. The associated density of states $n_{\mathrm{BP}}(E)$ and $n_{\mathrm{VCP}}(E)$ are given by inserting these V^c into (3.46). The results are plotted in Figure 4.

Finally, V_c can be determined from (3.52) selfconsistently. This is the coherent potential approximation and also Klauder's fifth approximation. If we define

$$\Sigma = [-2(E - V_c)]^{-1/2} = i[2(E - V_c)]^{-1/2}, \tag{3.55}$$

then equation (3.46) can also be written

$$n(E)_{\mathrm{CPA}} = (1/\pi\varepsilon)\,\mathrm{Im}\,\Sigma \tag{3.56}$$

where equation (3.52) can be rewritten as Klauder's cubic equation

$$2E\Sigma^3 - 2(E + \varepsilon)\Sigma^2 + \Sigma - 1 = 0 \tag{3.57}$$

for Σ. This spectrum terminates when E becomes negative enough for Σ to have only real roots. An examination of Figure 4 shows that $E_{\min}$ is lower for n_{CPA} than it is for the cruder approximations n_{VC}, n_{BP} and n_{VCP}. This is a good part of the reason for a better fit of the cumulative distribution $N_{\mathrm{CPA}}(E)$ to the exact distribution than any of the other approximations (see Figure 5). For large ε, the lower bounds of the spectrum can be written as

$$E_{VC} > -\varepsilon, \qquad E_{BP} > -\varepsilon - (\varepsilon/2)^{1/2},$$
$$E_{VCP} > -\varepsilon - (\varepsilon^2/2)^{1/3}, \qquad E_{CPA} > -\varepsilon - (3/2)\varepsilon^{2/3}, \tag{3.58}$$

where terms of lower order in ε have been omitted.

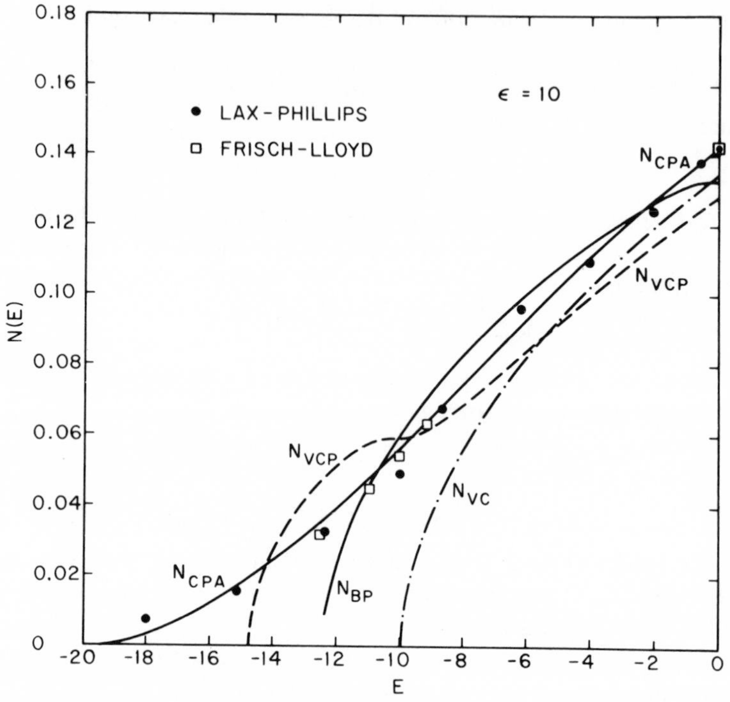

FIGURE 5

Integrated density of states $N(E)$ in a one-dimensional impurity band at high concentration $\varepsilon = n/K_0 = 10$. See Figure 4 for labelling of curves in corresponding differential density case. Circles are exact (Monte Carlo) results of Lax and Phillips [**1958**] and squares are corresponding exact (Markoff) results of Frisch and Lloyd [**1960**]. The VCP and CPA results are identical to the fourth and fifth approximations of Klauder [**1961**].

A formula asymptotically correct in the tail is found in the next section to be

$$n_{AS}(E) = \frac{2K^2}{\pi\varepsilon^2}\exp\left[-\frac{2}{3}\frac{K^3}{\varepsilon}\right], \tag{3.59}$$

where

$$K^2 = -2(E + \varepsilon). \tag{3.60}$$

Thus it is seen that the tail below the virtual crystal band edge is indeed of order $\varepsilon^{2/3}$ in extent. Moreover, the fraction of states per impurity atom below the virtual crystal band edge has the form

$$N(-\varepsilon) = -C\varepsilon^{-2/3}. \tag{3.61}$$

This was first observed by Lax and Phillips [**1958**]. A derivation by Frisch and Lloyd [**1960**] led to the exact value $C = 0.2532$.

The relationship

$$N(E) = \int_{E_{\min}}^{E} n(E)\, dE \tag{3.62}$$

leads to

$$N_{VC}(E) = (1/\pi\varepsilon)[2(E + \varepsilon)]^{1/2} \tag{3.63}$$

and

$$N_{VC}(0) = (1/\pi)(2/\varepsilon)^{1/2}. \tag{3.64}$$

This result is undoubtedly correct for large ε for the exact solution, since the extra contribution of the band tail of order $(1/\varepsilon)^{2/3}$ is smaller. But the tail is not negligible for reasonable size ε.

E. *The quasicrystalline approximation.* An alternate procedure for dealing with the multiple scattering equations (3.12) and (3.13) is to average them. The equation for $\langle \Psi \rangle$ is then found to depend on $\langle \Psi^j \rangle_j$ where the subscript j implies that scatterer j is held fixed. Similarly, the equation for $\langle \Psi^j \rangle_j$ will depend on $\langle \Psi^i \rangle_{ij}$ in which two scatterers are held fixed. The simplest truncation of these equations

$$\langle \Psi^j \rangle_j \approx \langle \overline{\Psi} \rangle \tag{3.65}$$

will be referred to as the effective field approximation (EFA). This approximation was first introduced by Foldy [**1945**] in treating isotropic scatterers and later used by Lax [**1951**] for anisotropic scatterers. A more sophisticated approximation introduced by Lax [**1952**],

$$\langle \Psi^i \rangle_{ij} \approx \langle \Psi^i \rangle_i, \tag{3.66}$$

was called the quasicrystalline approximation (QCA) because in the crystalline case holding j fixed as well as i does not change the effective field since all scatterers are already fixed. These approximations were used by Lax [**1952**] in connection with bare propagators. We have shown, however, that the form of our equations (3.27), (3.28) with modified propagators is identical to those (3.12) and (3.13) with unmodified propagators. Thus the results of Lax [**1952**] for the QCA with bare propagators can be immediately applied to the case of modified propagators just by renaming the unperturbed Hamiltonian (from H_0 to H) and the transition operator (from t to T), etc. This will be demonstrated briefly in Appendix B. A detailed treatment has been given by Gyorffy [**1970**].

A recent excellent review of the electron theory of liquid metals by Schwartz and Ehrenreich [**1971**] compares the CPA, i.e., the EFA with modified propagators

$$\langle G^j \rangle_j = \langle G \rangle = G_c \tag{3.67}$$

with the QCA with unmodified propagators. It is clear that the proper procedure is to use the QCA with modified propagators.

We cannot test the QCA (MP) (with modified propagators) on our one-dimensional example because in the purely random case, we (and Gyorffy) show the QCAMP reduces precisely to the CPA.

F. CPA *for alloys.* Our previous discussion of the coherent potential approximation for homogeneous systems has to be modified in the case of alloys. The coherent potential will not be diagonal in k space but will be periodic in ordinary space. Thus we can write

$$V_c = \sum_j v_c(\boldsymbol{r} - \boldsymbol{j}), \tag{3.68}$$

$$V_j = v(\boldsymbol{r} - \boldsymbol{j}) - v_c(\boldsymbol{r} - \boldsymbol{j}), \tag{3.69}$$

where $v(\boldsymbol{r} - \boldsymbol{j})$ takes one of two values $v_A(\boldsymbol{r} - \boldsymbol{j})$ or $v_B(\boldsymbol{r} - \boldsymbol{j})$ depending on whether an A or a B atom occupies site $\boldsymbol{j}$. Each kind of atom has its own T matrix, e.g.,

$$T_A = v_A - v_c + (v_A - v_c)e^{-1}T_A, \tag{3.70}$$

where

$$e = E - H_0 - V_c \tag{3.71}$$

with a similar equation for T_B. The coherent potential approximation (3.42) reduces in the alloy case with concentrations C_A and C_B to

$$C_A T_A + C_B T_B = 0 \tag{3.72}$$

as a condition for the selfconsistent determination of V_c with the result

$$v_c = C_A v_A + C_B v_B - (v_A - v_c)e^{-1}(v_B - v_c). \tag{3.73}$$

Here C_A and C_B are the concentrations of atoms A and B respectively. This condition (3.72) is not identical to using an effective t matrix which is the weighted average of the bare t_A and t_B matrices, a procedure suggested by Lax [**1951**] and Beeby [**1964**], now referred to as the average t matrix approximation ATA.

A procedure based on the selfconsistent condition (3.72) was suggested to D. W. Taylor who was already working on the lattice vibration problem and to P. Soven, whom I interested in the corresponding electronic energy band problem in alloys. Taylor [**1967**] was able to apply the method without difficulty and obtained results in good agreement with three-dimensional Monte Carlo calculations of Payton and Visscher [**1967**] except for a smoothing over the states in the impurity band.

Soven [**1966**] attempted a calculation of the electronic structure of beta brass. The difficulties of a selfconsistent approximation led him to try the average *t* matrix approximation (ATA). Later, at my suggestion, he used a one-dimensional model of an alloy to compare the ATA with the CPA. Figures 6 and 7 from Soven [**1967**] demonstrate the superiority of the CPA to the ATA. Moreover, the CPA was found to agree well with the exact results except near band edges, as Taylor had found in the lattice vibration case.

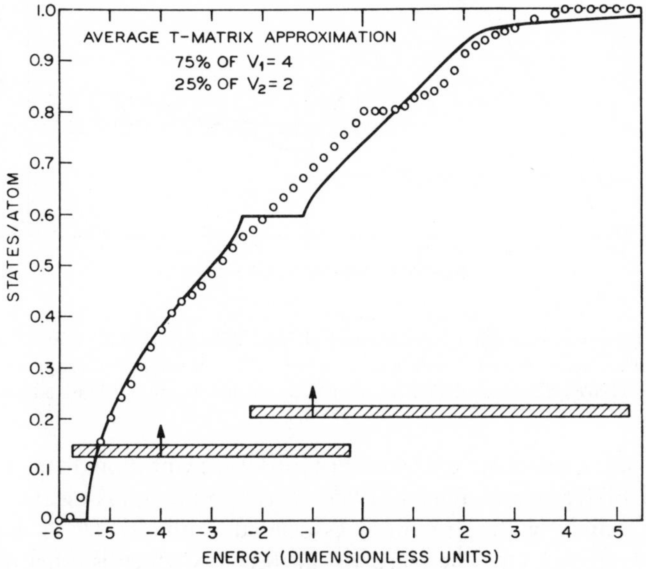

FIGURE 6

Integrated density of states in a one-dimensional 75%–25% binary alloy of delta function potentials. The circles are exact results and the curve is the average *t* matrix approximation. This figure is from Soven [**1967**].

Soven [**1969**] therefore applied the CPA to a three-dimensional tight binding model of an alloy. He also estimated pair corrections to the CPA. The smallness of these pair corrections, and the clarity of Soven's exposition has undoubtedly popularized the CPA.

It should be mentioned that Yonezawa [**1968**], and Yonezawa and Matsubara [**1966**] using a method due to Matsubara and Toyozawa [**1961**] have arrived at the CPA by a systematic diagram summation technique which includes corrections due to exclusion effects. Their simplest approximation, Yonezawa [**1968**], agrees with the Taylor [**1967**] CPA treatment of alloys and Yonezawa's [**1964**] treatment of the liquid metal case agrees with Klauder's approximation five, i.e. the CPA result (3.56).

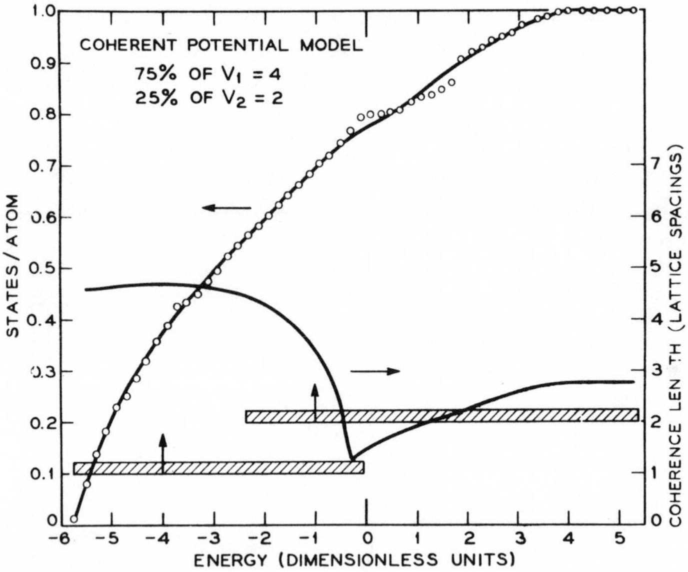

FIGURE 7

Integrated density of states in a one-dimensional 75%–25% binary alloy of delta function potentials. The circles are exact results and the curve is the coherent potential approximation (from Soven [**1967**]). The fit to the CPA is clearly better than the fit in Figure 6 to the ATA.

The Monte Carlo work of Dean [**1961**] on lattice vibrations in one dimension and of Dean and Bacon [**1965**] in two dimensions clearly shows impurity bands with structures associated with *ABABAB* clusters, *ABBABB* clusters, etc. These are lost in the CPA which is often referred to as a single site approximation. Generalizations of the CPA to n sites have been made by Nickel and Krumhansl [**1971**] and by Leath [**1972**].

When the potential in an alloy (or liquid metal) consists in a sum of nonoverlapping potentials, a simplification should result since each scattering can be regarded as taking place in a vacuum and should involve only "on-energy-shell" components of the scattering matrix (Goldberger and Watson [**1964**]). This is tacitly assumed in writing (3.3) since Φ_a should obey $H_0\Phi_a = E\Phi_a$, i.e. E and the momentum $\boldsymbol{a}$ are not to be regarded as independent. However, the t matrix equation (3.8) which results is general enough to solve for scattering from a wave such that $E \neq \hbar^2(\boldsymbol{a})^2/2m$.

The on-energy-shell information is most readily prescribed in terms of the scattering phase shifts in an angular momentum representation. The relation between the $\boldsymbol{k}$ and angular momentum representations is made clear in Beeby [**1964**] and the connection with the Korringa [**1947**] and Kohn-Rostoker [**1954**] formulation of electronic energy bands is worked

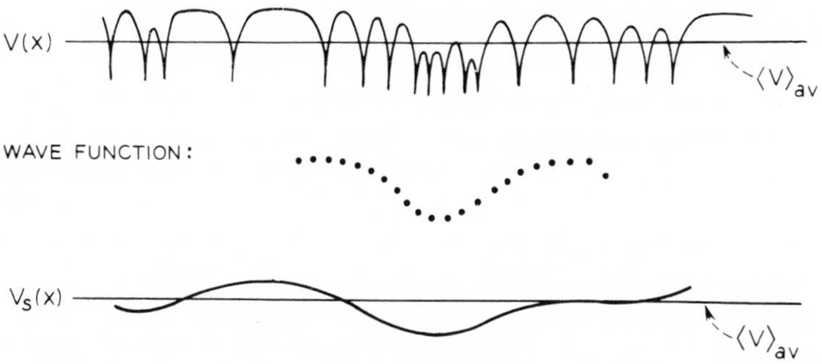

FIGURE 8

A pictorial description of the random potential $V(x)$ produced by a high density of impurities, of the localized wave-function trapped in a high concentration region of impurity centers, and of the smoothed potential $V_s(x)$ which is effective in the trapping. The smoothed potential is the original potential convoluted with the square of the localized wave function.

out by Beeby and Edwards [**1963**] and Ziman [**1966**] who also establishes the connection with the Lax [**1952**] multiple scattering formalism.

Soven [**1966**] made use of the dependence on on-energy-shell information by constructing an energy dependent shell potential whose scattering phase shifts are the same as those of the original atoms. Shiba [**1971**] gave a partial and Gyorffy [**1972**] a complete derivation of the selfconsistency condition of the CPA without the need for an auxiliary shell potential.

4. **Minimum counting methods.** The failure (see Figures 4 and 5) of the CPA to yield the deep tails in the density of states suggests that the averaging procedures of the CPA must be avoided. The electron must be allowed in a nonperturbative manner to follow fluctuations in the potential $V(\boldsymbol{x})$. But the Heisenberg uncertainty principle embedded in the Schrodinger equation prevents the electron from following very short range fluctuations (about the average potential $\langle V\rangle_{\text{av}}$). Instead, the electron responds to a smoothed version, $V_s(\boldsymbol{x})$, of the original potential (see Figure 8). (For detailed justification of the arguments presented here see Halperin and Lax [**1966**], [**1967**] and Zittartz and Langer [**1966**].)

The states deep in the tail are localized states of an electron trapped in a deep fluctuation of $V_s(\boldsymbol{x})$. If $V_s(\boldsymbol{x})$ possesses a minimum at $\boldsymbol{x}_0$, the localized wavefunction will have the form

$$\psi(\boldsymbol{x}) = f(\boldsymbol{x} - \boldsymbol{x}_0), \tag{4.1}$$

where $f(\boldsymbol{x})$ is to be determined later and may depend on the energy E. The energy eigenvalue associated with this wave function is

$$E(\boldsymbol{x}_0) = \langle f(\boldsymbol{x} - \boldsymbol{x}_0)| -\tfrac{1}{2}\nabla^2 + V(\boldsymbol{x})|f(\boldsymbol{x} - \boldsymbol{x}_0)\rangle = T + V_s(\boldsymbol{x}_0), \tag{4.2}$$

where T is the average kinetic energy and $V_s(\boldsymbol{x}_0)$ is the average potential energy which will also be used as the smoothed version of the original potential energy. In general, by the variational principle, $E(\boldsymbol{x}_0)$ will be greater than the actual energy of the lowest energy eigenstate E_i localized in the vicinity of $\boldsymbol{x}_0$. The error can be minimized by optimizing with respect to $f(\boldsymbol{x} - \boldsymbol{x}_0)$. Assuming the shape of $f(\boldsymbol{x})$ is already optimal, this is accomplished by varying $\boldsymbol{x}_0$ to minimize $E(\boldsymbol{x}_0)$. *Thus each local minimum in* $E(\boldsymbol{x}_0)$ *can be associated with a bound state* at the energy $E(\boldsymbol{x}_0)_{\min}$. (We assume, here, that we are concerned with states deep enough in the tail for these minima to occur infrequently so that the wave functions at different minima do not overlap.) The density of states *per unit volume* Ω can therefore be approximated by

$$\rho_f(E) = \frac{1}{\Omega\, dE} \cdot [\text{Number of local minima in } E(\boldsymbol{x}_0) \text{ in the interval } E, E + dE]. \tag{4.3}$$

For E deep in the tail, a stationary point $\boldsymbol{y}_i$ such that

$$\nabla V_s(\boldsymbol{y}_i) = 0 \tag{4.4}$$

is overwhelmingly likely to be a minimum point. Thus we must count the zeroes of $\nabla V_s(\boldsymbol{y})$. For simplicity, we first note that the number of zeroes of a one-dimensional function $h(y)$ in the interval $[a, b]$ is given by

$$\int_a^b \delta(h(y))|dh/dy|\, dy,$$

where the Jacobian dh/dy has been inserted so that each zero gives unit contribution to the integral. This is equivalent to the statement that

$$\sum_j \delta(y - y_j) = \delta(h(y))|dh(y)/dy|,$$

where the y_j are all the zeroes of $h(y)$. A generalization of this statement to three dimensions

$$\sum_j \delta(\boldsymbol{y} - \boldsymbol{y}_j) = \delta(\nabla V_s(\boldsymbol{y}))|\det \nabla\nabla V_s(\boldsymbol{y})| \tag{4.5}$$

makes use of the three-dimensional Jacobian. If we wish not merely the stationary points $\boldsymbol{y}_j$ obeying (4.4) but only those at energy

$$E(\boldsymbol{y}_j) = T + V_s(\boldsymbol{y}_j) = E,$$

the density of states is given by

$$\rho_f(E) = \frac{1}{\Omega}\int dy\, \langle \delta[E - T - V_s(y)]\delta(\nabla V_s(y)) \det \nabla\nabla V_s(y)\rangle, \tag{4.6}$$

where absolute value signs around the determinant are omitted since the determinant is positive with high probability at deep stationary points in the tail. Moreover, the ensemble average yields a result independent of $\boldsymbol{y}$, so that the integration over $\boldsymbol{y}$ merely cancels the volume Ω in the denominator.

Since energies are overestimated by $E(\boldsymbol{x}_0)$, the density of states deep in the tail are underestimated. The best choice for f at each energy E is clearly that which maximizes $\rho_f(E)$. A first approximation to the density of states can then be written

$$\rho_1(E) = \max_f [\rho_f(E)]. \tag{4.7}$$

This result applies regardless of the statistics assumed for the potential $V(\boldsymbol{x})$. Extensive calculations have been made by Halperin and Lax [**1966**] for the case in which the statistics are Gaussian, with the first moment of $V(\boldsymbol{x})$ shifted to zero, and the second moment given by

$$\langle V(\boldsymbol{x})V(\boldsymbol{x}')\rangle = \xi W(\boldsymbol{x} - \boldsymbol{x}'). \tag{4.8}$$

The complete statistics of $V(\boldsymbol{x})$ are defined by $\langle V(\boldsymbol{x})\rangle = 0$ and equation (4.8). It is then possible to make an essentially exact evaluation of $\rho_f(E)$ with the result

$$\rho_f(E) = \frac{[\sigma_1(T - E)]^d}{(2\pi\xi)^{1/2(d+1)}(\sigma_0)^{(2d+1)}} \exp[-\Gamma/(2\xi)], \tag{4.9}$$

where

$$\Gamma = \frac{(E - T)^2}{(\sigma_0)^2} = \frac{[E - \int f(\boldsymbol{x})(-\frac{1}{2}\nabla^2) f(\boldsymbol{x})\, d\boldsymbol{x}]^2}{\int f^2(\boldsymbol{x}) f^2(\boldsymbol{x}') W(\boldsymbol{x} - \boldsymbol{x}')\, d\boldsymbol{x}\, d\boldsymbol{x}'}, \tag{4.10}$$

$$(\sigma_1)^{2d} = \det \langle \nabla V_s(\boldsymbol{x}) \nabla V_s(\boldsymbol{x})\rangle, \tag{4.11}$$

$(\sigma_0)^2$ is given by the denominator of (4.10) and d is the dimension of the space, e.g., $d = 1, 2$ or 3. The tail region is one in which $E^2 \gg \xi$. In this region, $\rho_f(E)$ is very sensitive to the exponent Γ. Maximizing $\rho_f(E)$ consists then in minimizing Γ. Introducing a Lagrange multiplier μ, application of the methods of variational calculus to (4.10) leads to

$$-\tfrac{1}{2}\nabla^2 f(\boldsymbol{x}) - \mu f(\boldsymbol{x}) \int f(\boldsymbol{x}')^2 W(\boldsymbol{x} - \boldsymbol{x}')\, d\boldsymbol{x}' = E f(\boldsymbol{x}), \tag{4.12}$$

a Hartree equation for a particle bound in its own selfconsistent field. In the present case, however, E is specified and μ regarded as the eigenvalue rather than vice-versa.

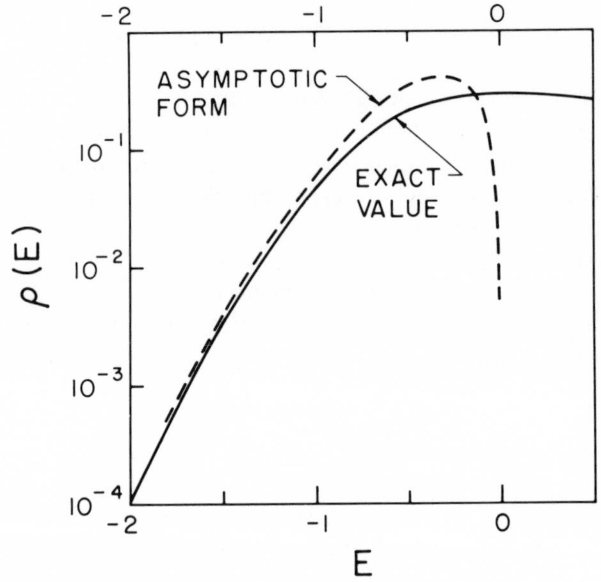

FIGURE 9

Differential density of states for a one-dimensional white noise model: $\langle V(x)\rangle = E_0$, $\langle \Delta V(x)\Delta V(x')\rangle = \varepsilon\delta(x - x')$. Comparison of the *exact* $\rho(E)$ with the asymptotic form of the exact $\rho(E)$, namely $\rho_{AS}(E)$. A universal curve is found by plotting $(2\varepsilon)^{1/3}\rho$ against $U = (E - E_0)/(2\varepsilon)^{2/3}$ and $(2\varepsilon)^{1/3}\rho_{AS} = (-8U/\pi)\exp[-c(-U)^{3/2}]$ where $c = 8(2)^{1/2}/3$. The point $E = -1$ on the absissa actually represents $U = -1$ or $E = E_0 - (2\varepsilon)^{2/3}$. Units in which $\hbar = m = 1$ have been used.

The accuracy of this method can be tested in the one-dimensional delta-correlated case of §3 for which

$$\langle V(x)V(x')\rangle = \varepsilon\delta(x - x') \tag{4.13}$$

since the resulting local nonlinear equation

$$-\tfrac{1}{2}f''(x) - \tfrac{1}{2}\mu f(x)^3 = Ef(x) \tag{4.14}$$

has an exact solution

$$f(x) = (K/2)^{1/2}\,\mathrm{sech}\,Kx, \tag{4.15}$$

where K is defined by

$$E = -\tfrac{1}{2}K^2 - \varepsilon \tag{4.16}$$

and the eigenvalue μ is found to be $4K$. The resulting density of states

$$\rho_1(E) = \frac{1}{(5)^{1/2}}\frac{2K^2}{\pi\varepsilon}\exp\left(-\frac{2}{3}\frac{K^3}{\varepsilon}\right) \tag{4.17}$$

is in excellent agreement with the correct asymptotic ($E \to -\infty$) form of

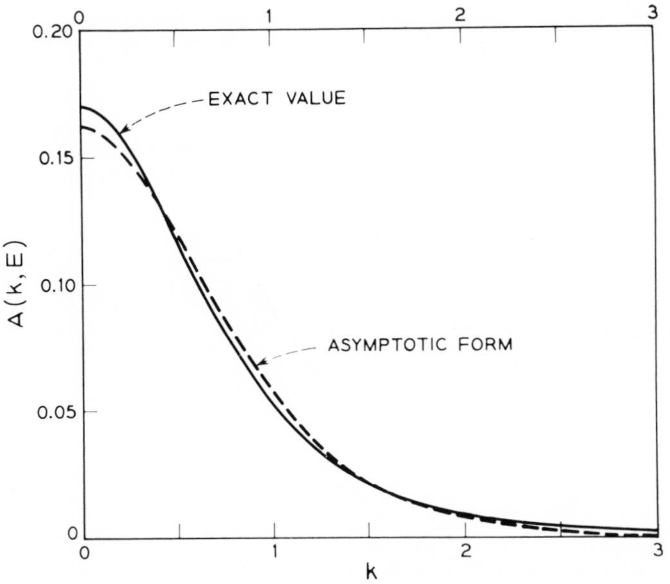

FIGURE 10

Momentum dependence of spectral density for one-dimensional white noise model (see Figure 9 caption), at energy $E = E_0 - (2\varepsilon)^{2/3}$. Comparison of $A(k, E)$ and its asymptotic form $|F(k)|^2\rho(E)$ where $|F(k)|^2 = (\pi^2/2K)\,\mathrm{sech}^2(\pi k/2K)$ and $E - E_0 = -(2\varepsilon)^{2/3} = -\frac{1}{2}K^2$ at the energy plotted. We plot $(2\varepsilon)^{2/3}A(k, E)$ versus $k/(2\varepsilon)^{1/3}$. We have assumed $\hbar = m = 1$.

the density of states: The third power of K in the exponent is correct, the coefficient $\frac{2}{3}$ is correct, the second power of K in the prefactor is correct. The numerical coefficient is too small by a factor $(5)^{1/2}$. [The exact result is obtainable from Frisch and Lloyd [**1960**], Halperin [**1965**] and §4 of Lax [**1966**] and the correct asymptotic form is given in (3.59) with a different normalization $n(E) = \rho(E)/\varepsilon$.]

Zittartz and Langer [**1966**] and Halperin and Lax [**1967**] consider second-order corrections to the above described minimum counting approach arising from differences between the actual well shape and the typical potential well associated with the typical wave function $f(\mathbf{x})$. It is shown by Halperin and Lax [**1967**] that after second-order corrections are applied, the correct asymptotic distribution $\rho_{AS}(E)$ is obtained, both in three dimensions and for our one-dimensional white noise example.

The asymptotic distribution (3.59) is not identical (see Figure 9) to the exact distribution except deep in the tail because only the ground states in each will have been included in $\rho_f(E)$. Hence the big discrepancy near $E = 0$ where the states are likely to be free rather than bound. Indeed, the agreement between asymptotic and exact $\rho(E)$ sets in fairly quickly below the virtual crystal band edge.

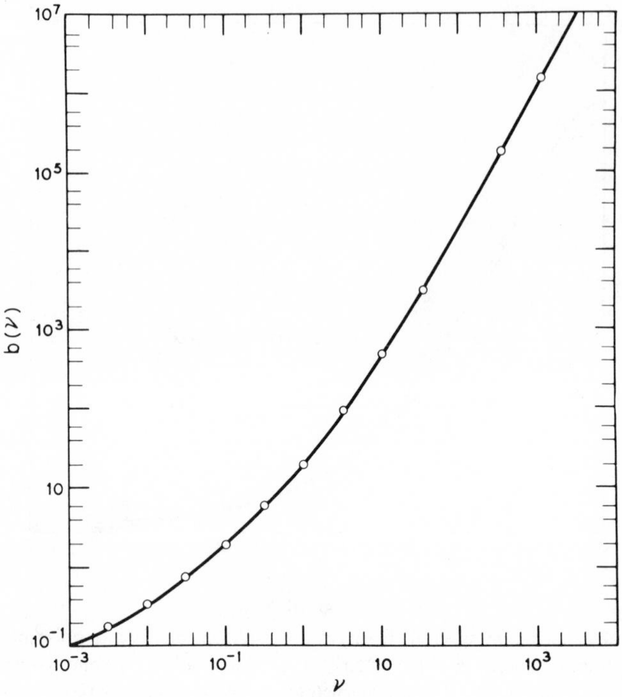

FIGURE 11

The density of states deep in the tail of an impurity band for a density of impurities high enough to use Gaussian statistics has the form

$$\rho_1(E) = ((QE_Q)^3/\xi^2)a(v)\exp[-(E_Q^2/2\xi)b(v)],$$

where $v = (E_0 - E)/E_Q$ is an energy measured relative to the mean potential energy E_0 in units $E_Q = \hbar^2 Q^2/(2m^*)$, where Q is the inverse screening radius. The parameter

$$\xi = \frac{2\pi}{Q}\frac{e^4}{\varepsilon_0^2}\sum_a \bar{n}_a Z_a^2,$$

where $\bar{n}_a$ is the concentration of impurities with charge $Z_a e$, ε_0 is the dielectric constant, and m^* is the effective mass of the band from which the impurity band is formed. The figure shows the universal function $b(v)$ which varies from $v^{1/2}$ at small v to v^2 at large v.

A test more sensitive to the shape of the wave function $f(\boldsymbol{x})$ at energy E is the spectral density which is proportional to the square of the Fourier transform $F(\boldsymbol{k})$ of $f(\boldsymbol{x})$ and the density of states $\rho(E)$:

$$A(\boldsymbol{k}, E) = |F(\boldsymbol{k})|^2 \rho(E). \tag{4.18}$$

Figure 10 displays the excellent agreement between this choice of spectral density and the exact spectral density.

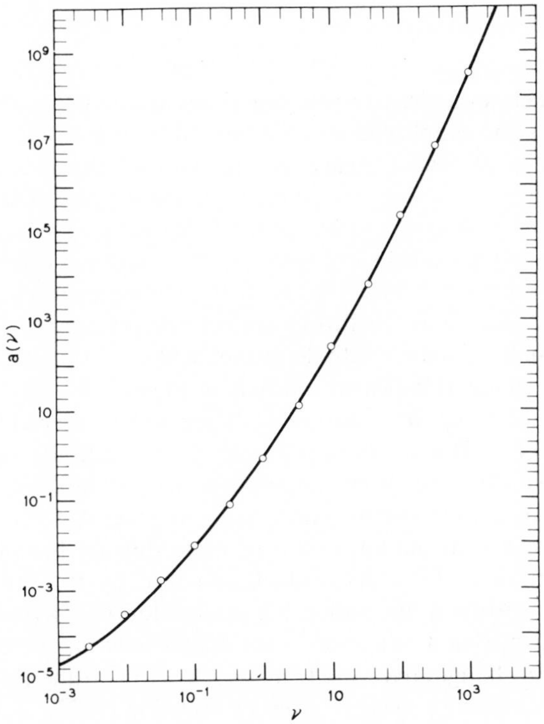

FIGURE 12

The universal prefactor $a(\nu)$ in the density of states of equation (4.21) is plotted against $\nu = (E_0 - E)/E_Q$ (see Figure 11 caption).

Having adequately tested the method in one dimension, Halperin and Lax [**1966**] solve numerically for $f(\mathbf{x})$ in three dimensions for the case of screened Coulomb interactions which lead to a correlation of the form

$$\langle V(\mathbf{x})V(\mathbf{x}')\rangle = \xi \exp(-Q|\mathbf{x} - \mathbf{x}'|), \tag{4.19}$$

$$\xi = \frac{2\pi}{Q}\frac{e^4}{\varepsilon_0^2}\sum_a \bar{n}_a Z_a^2, \tag{4.20}$$

where $1/Q$ is the screening radius, ε_0 is the dielectric constant of the pure semiconductor and $Z_a e$ is the charge of the a-type impurity whose mean density is $\bar{n}_a$. The density of states is shown to be of the form

$$\rho_1(E) = ((QE_Q)^3/\xi^2)a(\nu)\exp[-(E_Q^2/2\xi)b(\nu)], \tag{4.21}$$

where

$$E_Q = \hbar^2 Q^2/2m^* \tag{4.22}$$

is the energy unit and

$$\nu = (E_0 - E)/E_Q > 0 \tag{4.23}$$

is the energy below the virtual crystal band edge in units of E_Q. The universal exponent $b(\nu)$ and amplitude $a(\nu)$ are plotted in Figures 11 and 12, respectively. These plots cover many decades. It is clear that over one decade, $b(\nu)$ can be replaced by a straight line on the log-log plot. Thus $b(\nu) \sim \nu^n$. To see how the exponent n varies from $\frac{1}{2}$ to 2 over a wide range of ν, we show a plot of $d \log b/d \log \nu$ in Figure 13. This variation can be regarded as produced (at fixed ν) by a change in the scaling energy E_Q, i.e. by a change in material. Thus this theory can account for a range of behaviors from a Gaussian tail to tails with the slower $\exp(-\nu^{1/2})$ cut-off. The power law variation of $a(\nu)$ is indicated similarly in Figure 14.

The procedures up to equation (4.7) are valid independent of the statistics of $V(\boldsymbol{x})$. The subsequent results have made use of Gaussian statistics. Since $V(\boldsymbol{x})$ arises from a superposition of potentials it depends on the concentration fluctuations which have, in general, a Poisson rather than a Gaussian distribution. It is then more difficult to evaluate (4.6). The exponent in $\rho_f(E)$ is determined, however, by the probability of having a concentration fluctuation big enough for $T + V_s(y)$ to equal E. The requirement that a minimum occur affects only the pre-exponential factors and can be ignored. An estimate of $\rho_f(E)$ and the resulting self-consistent equations for $f(\boldsymbol{x})$ are given by Halperin [**1972**].

5. **Path integral formulation.** The reformulation of quantum mechanics via path integrals by Feynman [**1948**] (see also Feynman and Hibbs [**1965**]) has been exploited by Edwards [**1970**] to provide an alternative approach to the density of states and the conductivity problem. Edwards' description of his own work is rather terse but an expanded, improved version of Edwards' work is given by Economou, Cohen, Freed and Kirkpatrick [**1971**]. In view of this extensive discussion, we shall only comment briefly on Edwards' work here.

The key theorem of the Feynman technique is that the solution of the time dependent Green's function equation

$$[i\partial/\partial t + \tfrac{1}{2}\nabla^2 - V(\boldsymbol{r})]G(\boldsymbol{r}\boldsymbol{r}'t, V) = \delta(\boldsymbol{r} - \boldsymbol{r}')\delta(t) \tag{5.1}$$

can be written as the path integral

$$G(\boldsymbol{r}\boldsymbol{r}'t, V) = \int_{\boldsymbol{r}(0)=\boldsymbol{r}'}^{\boldsymbol{r}(t)=\boldsymbol{r}} D[\boldsymbol{r}(s)] \exp\left\{i\int_0^t ds[\tfrac{1}{2}\dot{\boldsymbol{r}}(s)^2 - V(\boldsymbol{r}(s))]\right\}. \tag{5.2}$$

Since the Green's function is the $\boldsymbol{r}, \boldsymbol{r}'$ matrix element of the operator

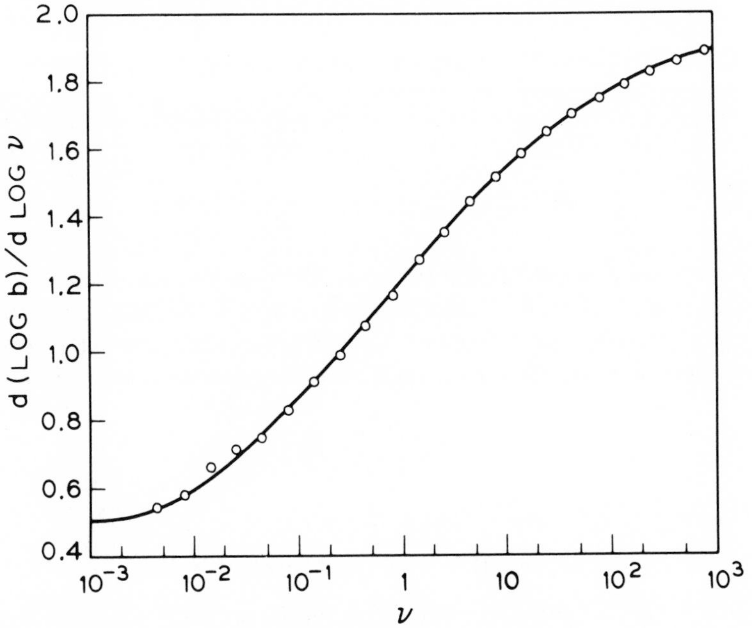

FIGURE 13
The logarithmic derivative $n = d \log b(\nu)/d \log \nu$ of the exponent $b(\nu)$ in the density of states is plotted and shown to vary smoothly from $n = \frac{1}{2}$ to 2.

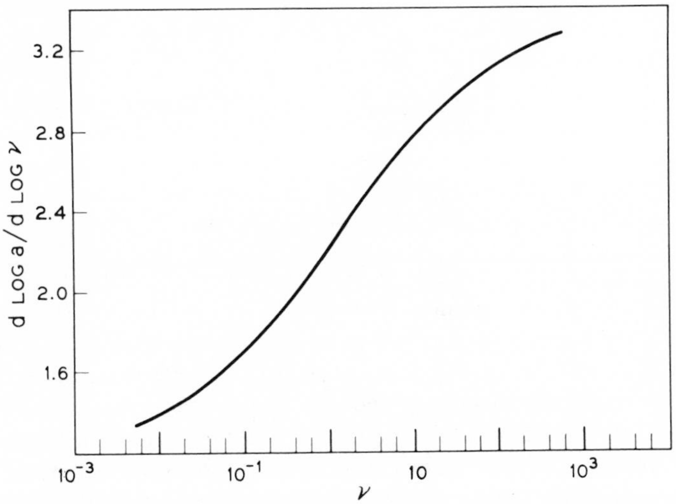

FIGURE 14
The logarithmic derivative $d \log a(\nu)/d \log \nu$ of the prefactor $a(\nu)$ in the density of states.

$\exp[-it(H_0 + V)]$, and the density of states (3.22) is the trace of

$$\langle \delta(E - H_0 - V) \rangle = \frac{1}{2\pi} \int_{-\infty}^{\infty} e^{iEt} \langle \exp[-it(H_0 + V)] \rangle \, dt, \tag{5.3}$$

we can write

$$n(E) = \int d\mathbf{r} \frac{1}{2\pi} \int_{-\infty}^{\infty} e^{iEt} \langle G(\mathbf{r}\mathbf{r}t, V) \rangle \, dt. \tag{5.4}$$

Economou et al. [**1971**] and Edwards [**1970**] go through a long derivation during which they assume the potential V is produced by a very high concentration of weak potentials. Under such circumstances, the central limit theorem guarantees that $V(\mathbf{r})$ is a *Gaussian random variable.* But then

$$g = \int_0^t ds \, V(\mathbf{r}(s)) \tag{5.5}$$

is Gaussian and, for any Gaussian variable g,

$$\langle \exp(-ig) \rangle = \exp(-i\langle g \rangle) \exp(-\tfrac{1}{2}\langle g^2 \rangle), \tag{5.6}$$

so that we obtain directly from (5.2) one of the principal equations of the Edwards theory,

$$\langle G(\mathbf{r}\mathbf{r}', t) \rangle = \int_{\mathbf{r}(0)=\mathbf{r}'}^{\mathbf{r}(t)=\mathbf{r}} D[\mathbf{r}(s)] \cdot \exp\left\{ \frac{1}{2} i \int_0^t ds [\dot{\mathbf{r}}(s)]^2 - \frac{1}{2} \int_0^t ds \int_0^t ds' \, W(\mathbf{r}(s) - \mathbf{r}(s')) \right\}, \tag{5.7}$$

where

$$\langle V(\mathbf{r}) V(\mathbf{r}') \rangle = W(\mathbf{r} - \mathbf{r}') \tag{5.8}$$

is the correlation of the potential fluctuations. The displacement invariance of the theory which requires that $\langle G(\mathbf{r}, \mathbf{r}', t) \rangle = G(\mathbf{r} - \mathbf{r}', t)$ does not provide a mechanism for any preferred origin, thus seeming to preclude localized states. The situation is analogous to the ferromagnetic phase transition in which symmetry requires the average magnetization to vanish. However, if a mean magnetization is assumed, and the magnetization is calculated selfconsistently it is found not to vanish below the transition temperature. To obtain this symmetry breaking field, the above-mentioned authors following Gelfand and Jaglom [**1960**] and Siegert [**1963**] introduce an external Gaussian random field ϕ. This is all unnecessary, since we already have the Gaussian potential $V(\boldsymbol{r})$. The average over $V(\boldsymbol{r})$ can be made explicit by writing

$$G(\boldsymbol{r}\boldsymbol{r}'t) = \langle G(\boldsymbol{r}\boldsymbol{r}'t, V)\rangle$$

(5.9)

$$= N \int D[V] \exp[-B(\boldsymbol{r}\boldsymbol{r}', tV)],$$

where

(5.10) $$B = -\ln G(\boldsymbol{r}\boldsymbol{r}'t, V) + \frac{1}{2}\int d\boldsymbol{r} \int d\boldsymbol{r}'\, V(\boldsymbol{r})W^{-1}(\boldsymbol{r} - \boldsymbol{r}')V(\boldsymbol{r}'),$$

and $W^{-1}(\boldsymbol{r} - \boldsymbol{r}')$ is the inverse of $W(\boldsymbol{r} - \boldsymbol{r}')$ regarded as a matrix on the indices $\boldsymbol{r}$ and $\boldsymbol{r}'$. The second part of B is chosen to yield the correct second moment, (5.8), and N is the normalization factor

(5.11) $$N^{-1} = \exp\left[-\frac{1}{2}\int\int V(\boldsymbol{r})W^{-1}(\boldsymbol{r} - \boldsymbol{r}')V(\boldsymbol{r}')\, d\boldsymbol{r}\, d\boldsymbol{r}'\right] D[V].$$

The most likely potential V, which will supply the symmetry breaking arises naturally when one tries to evaluate the integral in (5.9) by expanding B about its minimum which is determined by the condition

(5.12) $$\frac{\delta B}{\delta V(\boldsymbol{r}'')} = -\frac{1}{G(\boldsymbol{r}, \boldsymbol{r}')}\frac{\delta G(\boldsymbol{r}, \boldsymbol{r}')}{\delta V(\boldsymbol{r}'')} + \int W^{-1}(\boldsymbol{r}'' - \boldsymbol{r}''')V(\boldsymbol{r}''')\, d\boldsymbol{r}''' = 0.$$

The functional derivative is simplified by using

(5.13) $$GG^{-1} = 1, \qquad \frac{\delta G}{\delta V}G^{-1} + G\frac{\delta G^{-1}}{\delta V} = 0$$

so that

(5.14) $$\frac{\delta G(\boldsymbol{r}\boldsymbol{r}'t0)}{\delta V(\boldsymbol{r}'')} = -\int G(\boldsymbol{r}\boldsymbol{r}_1)\, d\boldsymbol{r}_1\, dt_1 \frac{\delta G^{-1}(\boldsymbol{r}_1\boldsymbol{r}_2)}{\delta V(\boldsymbol{r}'')} d\boldsymbol{r}_2\, dt_2\, G(\boldsymbol{r}_2\boldsymbol{r}'),$$

where

(5.15) $$G^{-1}(\boldsymbol{r}_1\boldsymbol{r}_2t_1t_2, V) = [i\partial/\partial t + \tfrac{1}{2}\nabla_1^2 - V(\boldsymbol{r}_1)]\delta(\boldsymbol{r}_1 - \boldsymbol{r}_2)\delta(t_1 - t_2),$$

(5.16) $$\delta G^{-1}/\delta V(\boldsymbol{r}'') = -\delta(\boldsymbol{r}_1 - \boldsymbol{r}'')\delta(\boldsymbol{r}_1 - \boldsymbol{r}_2)\delta(t_1 - t_2).$$

Combining the last three equations, we get

$$\frac{\delta G(\mathbf{r}\mathbf{r}', t0, V)}{\delta V(\mathbf{r}'')} = \int_0^t G(\mathbf{r}\mathbf{r}'', ts)\, ds\, G(\mathbf{r}''\mathbf{r}', s0) \tag{5.17}$$

where the dependence of G on V is understood.

Equation (5.12) can be solved explicitly for V by inverting the "matrix" W^{-1} with the result

$$V(\mathbf{r}, t) = \int W(\mathbf{r} - \mathbf{r}'')\, d\mathbf{r}'' \int_0^t ds \frac{G(\mathbf{r}\mathbf{r}'', ts, V)G(\mathbf{r}''\mathbf{r}', s0, V)}{G(\mathbf{r}\mathbf{r}', t0, V)}, \tag{5.18}$$

which displays explicitly the selfconsistent nature of V as well as its dependence on $\mathbf{r}'$. For the density of states (5.4), one can set $\mathbf{r}' = \mathbf{r}$. Calculation of the conductivity requires, however, the average of the product of two Green's functions (with $\mathbf{r} \neq \mathbf{r}'$). So far only qualitative results have been obtained for the conductivity. Detailed calculations are also required for the density of states. The expression (5.18) is promising, however, since it reduces, if one retains only the ground state contribution (to G) to a result proportional to the Halperin-Lax [**1966**], [**1967**] selfconsistent potential.

6. **The Anderson nondiffusion problem.** Anderson [**1958**] demonstrated that when electrons are subject to random potential fluctuations larger than a critical value (of the order of the bandwidth of the nondisordered crystal) all the electronic states acquire a local character: A charge at the origin does not diffuse to infinity but retains a finite probability of remaining at the origin even at infinite time. Anderson's work has been extended and illuminated by the mobility edge model of a random solid developed by Mott [**1961**]–[**1970**] and his collaborators, Mott and Twose [**1961**], Mott and Allgaier [**1967**], Mott and Davis [**1968**], Cutler and Mott [**1969**], and Austin and Mott [**1969**]. Further insight and calculations have been provided by Cohen, Fritsche and Ovshinsky [**1969**], Economou, Kirkpatrick, Cohen, and Eggarter [**1970**] and Cohen [**1970**] and by Thouless [**1972**]. The essence of the Mott-CFO model is that the presence of disorder takes the states of the perfect crystal and produces band tails of localized nature (see Figure 15). The states in the range $E_c < E < E_c'$ are extended d.c. current carrying states. As the disorder increases, the energy interval $E_c' - E_c$ decreases until the Anderson transition is reached, at which point all states are localized.

The specific model for which Anderson carried out his calculations is described by a matrix

$$H_{mn} = \varepsilon_m \delta_{mn} + V_{mn} \tag{6.1}$$

in which the V_{mn} are nonrandom coupling matrix elements between sites m and n, and the ε_m are a set of uncorrelated randomly distributed energies.

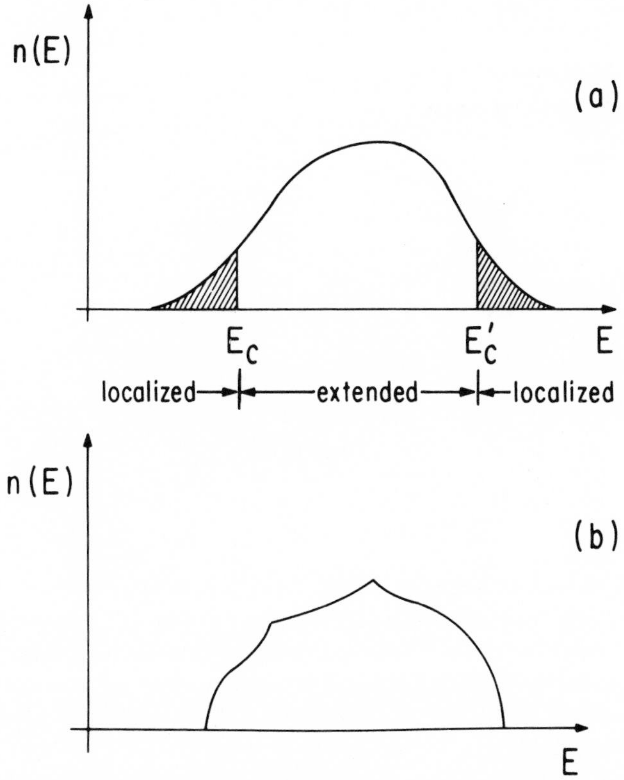

FIGURE 15
The density of states in a perfect lattice (b) acquires tails in a discordered system (a). The states in the interval $E_c < E < E_c'$ are extended. The remainder are localized. As disorder is increased, the interval $E_c' - E_c$ decreases.

The probability amplitude for an electron starting at site 0 to remain at site 0 is given by

$$(e^{-iHt})_{00} = \lim_{\eta \to 0^+} \int_{-\infty + i\eta}^{\infty + i\eta} G_0(E)\, e^{-iEt}\, dE/2\pi i, \tag{6.2}$$

where

$$G_0(E) = \left(\frac{1}{E - H}\right)_{00} = \frac{1}{E - \varepsilon_0 - \Delta(E)}, \tag{6.3}$$

where H now stands for the total Hamiltonian (previously $H_0 + V$) and $\Delta(E)$ is the "self-energy" of site 0. The density of states per unit volume can be written

$$\rho_0(E) = \langle \delta(E - H)_{00} \rangle = (1/\pi) \lim_{s \to 0^+} \operatorname{Im}\langle G_0(E - is) \rangle. \tag{6.4}$$

To determine whether the electron is trapped in the vicinity of the origin, we ask for the probability at infinite time that an electron starting at site 0 remains at site 0:

$$\begin{aligned} P_{00} &= \lim_{s \to 0^+} s \int_0^\infty e^{-st}\, dt |(e^{-iHt})_{00}|^2 \\ &= \int_{-\infty}^{\infty} f_0(E)\, dE, \end{aligned} \tag{6.5}$$

$$\begin{aligned} f_0(E) &= (s/\pi) G_0(E - is) G_0(E + is) \\ &= \frac{1}{\pi} \frac{s[G_0(E - is) - G_0(E + is)]}{2is - [\Delta(E + is) - \Delta(E - is)]} \\ &= \rho_0(E) \frac{1}{1 - (\Delta(E + is) - \Delta(E - is))/2is}, \end{aligned} \tag{6.6}$$

where the limit s approaching 0^+ is understood in (6.6).

If $P_{00} = 0$ the electron definitely escapes to infinity, indicating the presence of extended states. Since $f_0(E)$, in its first form, is manifestly positive definite, if $P_{00} = 0$, $f_0(E) = 0$ for all E so that all states are extended. If $\rho_0(E) \neq 0$ but $f_0(E) = 0$ for some E, then states at E exist and are extended. If $\Delta(E)$ has a branch cut in the interval $[E_1, E_2]$, then

$$\lim_{s \to 0} [\Delta(E + is) - \Delta(E - is)] \neq 0 \tag{6.7}$$

so that the denominator in (6.6) becomes infinite and $f_0(E)$ vanishes along the branch cut. Thus, states which exist in the energy interval of a branch cut are extended.

From (6.3), we can conclude that $G_0(E)$ and $\Delta_0(E)$ have branch cuts at the same place so that the extended states exist on the branch cuts of $G_0(E)$. Conversely, if $G_0(E)$ has a pole at E_j, then $f_0(E_j) = \rho_0(E_j) > 0$ and the state at E_j is localized.

Since $\Delta(E)$ is analytic at points E_j at which $G_0(E)$ has poles (and the states are localized) and is nonanalytic (with a branch cut) at energies at which the states are extended, the convergence (divergence) of the perturbation series for $\Delta(E)$ at E determines the localization (nonlocalization) of the states at energy E. A renormalized perturbation series (RPS) with nonrepeating indices is used by Economou et al. [**1971**]. This series is attributed to Watson [**1957**] but is originally due to Feenberg [**1948**], and is discussed in detail in Morse and Feshbach [**1953**].

The analysis of the convergence of the RPS for $\Delta(E)$ is complicated for two reasons: (1) It is not sufficient to consider $\langle \Delta(E) \rangle$ since the analytical arguments separating localized from nonlocalized states get blurred if an ensemble average of $\Delta(E)$ is taken. Convergence except for a set of measure zero of the RPS is adequate; however, as is an ensemble average of $f_0(E)$. (2) The work of Economou and Cohen [**1972**] and of Anderson [**1958**] differ in their assumptions concerning the statistical independence of terms of different order, in particular the relative signs of such terms.

It is of interest to note that Lloyd [**1969**] was able to find an *exact* solution for $\langle G \rangle$ when ε_i have a Lorentz distribution and Brouers [**1970**] was able to find a corresponding exact solution for $\langle \Delta(E) \rangle$. Nevertheless, both authors *incorrectly* concluded that localized states do not exist because they assumed that the existence of $\mathrm{Im}\langle G \rangle$ or $\mathrm{Im}\langle \Delta \rangle$ was sufficient to yield extended states whereas the probability distribution of Δ or G is needed to arrive at definitive conclusions.

The arguments against localization by Lloyd [**1969**], Brouers [**1970**] and Ziman [**1969**] were refuted by Anderson [**1970**], Mott [**1970**] and Thouless [**1970**]. In spite of Thouless' more careful mathematical analysis this is clearly a field in which the rigor of the professional mathematician can profitably be employed.

7. **Percolation theory.**

A. *Definition.* Percolation theory was originally introduced to describe a process in which a fluid soaks into a porous medium. A diffusion or transport description is appropriate if each scattering event is random, but memory of previous history is lost before the next event. A percolation description is appropriate when the principal source of randomness is not in the scattering event but in the scattering medium. If the latter varies from place to place but is static in time, then a particle after it jumps remembers where it came from. For example, the first jump from a representative point in a cubic lattice may have equal probability of being in any of six directions after averaging over the environment. If jumps are more likely from points of high potential to points of low potential, the second jump is less likely to be in the return direction than some other direction. (This remark is obvious in the limit in which jumps can only go downhill.)

It might appear that all the problems described, so far, in this paper are percolation problems in the sense that we start with a static potential V and later average over an ensemble of such media. This is not so for three reasons: (1) the Schrodinger equation does not describe probabilities, but probability amplitudes, and only in certain cases are approximations justified which can reduce the problem to one of probabilities; (2) nevertheless, quantum mechanics introduces probabilities into the individual scattering events so that a diffusion or transport description is often more appropriate than a percolation description (see the next section for such a description); (3) the conventional description of percolation theory, as reviewed excellently by Shante and Kirkpatrick [**1971**] and by Frisch and Hammersley [**1963**] describes a set of bonds which are open or closed with probability p (or a set of atomic sites which are open or closed with probability p). Thus, for a given configuration, a particular jump in percolation theory has probability one or zero, rather than the finite jump rate descriptive of our physical problems.

In spite of these objections, the direct current (d.c.) conductivity of a sample requires a finite probability of traversal of a path across an entire sample. If one can approximate a network with variable resistors by resistors that are zero (open) or infinite (closed), the associated percolation problem does the correct counting of the number of allowed paths of various lengths, and yields the probability of finding the necessary paths of infinite length for d.c. conductivity.

B. *Percolation on a lattice.* If p is the fraction of unblocked bonds (sites), $P_N(p)$ is the probability of wetting N atoms starting from a representative atom and

$$P(p) = \lim_{N\to\infty} P_N(p) \tag{7.1}$$

is the probability of wetting an infinite number of atoms. A superscript b or s is added if one wishes to specifically designate the bond or site problem.

A critical probability p_c is defined such that $P(p) = 0$ for all $p < p_c$. In other words, for $p < p_c$ only a finite number of atoms get wet. Hammersley [**1961**] has shown that the site and bond problems obey the inequality

$$P_N^s(p) \leqq P_N^b(p) \tag{7.2}$$

so that, if $P(p)$ is a monotonic increasing function of p,

$$p_c(s) \geqq p_c(b). \tag{7.3}$$

Monte Carlo calculations of $P(p)$ for both the site and bond problem for several two- and three-dimensional lattices carried out by Frisch, Hammersley and Welsh [**1962**] are displayed in Figure 16. A comparison of critical probabilities p_c calculated by Monte Carlo methods (Vyssotsky et al. [**1961**], Frisch et al. [**1961**], Dean [**1963**]) and series methods (Sykes and Essam [**1964a**], [**1964b**]) are displayed in Table III.

The form of the curves in Figure 16 suggests that near p_c

(7.4) $$P(p) = A(p - p_c)^k$$

where the critical exponent is shown by Shante and Kirkpatrick [**1971**] to be bounded by

(7.5) $$0 \leqq k \leqq (d - 1)/d,$$

where d is the dimensionality of the space.

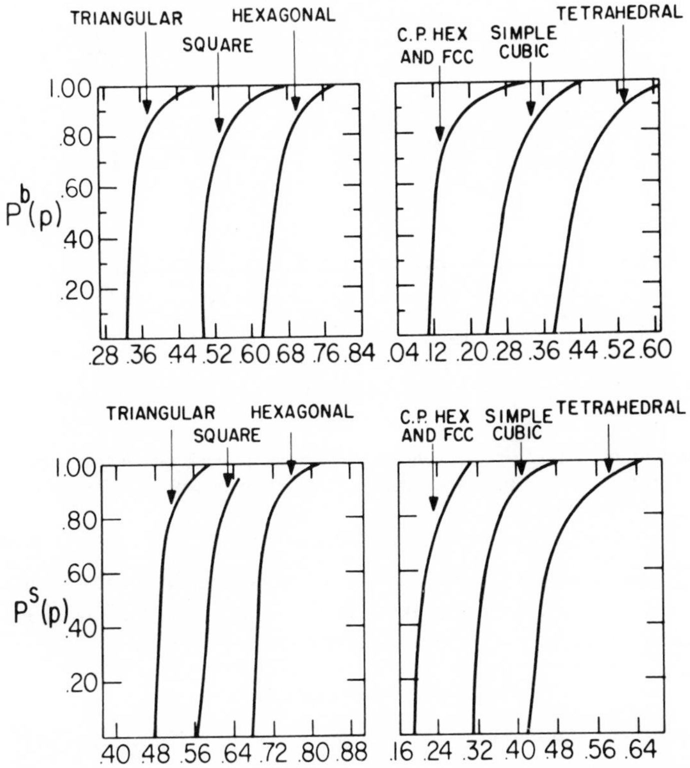

FIGURE 16

The probability $P^b(p)$ [or $P^s(p)$] of wetting an infinite number of atoms is plotted versus the fraction p of unblocked bonds [unblocked sites] for several two-dimensional and three-dimensional lattices. After Frisch, Hammersley and Welsh [**1962**].

C. *Relation of conductivity to percolation theory.* Several authors (Ziman [**1968**], Cohen [**1970**], Zallen and Scher [**1971**], Eggarter and Cohen [**1970**], [**1971**]) have stressed the intimate connection between a semiclassically calculated conductivity and the percolation probability. The mobility edge (which divides extended states that carry d.c. current

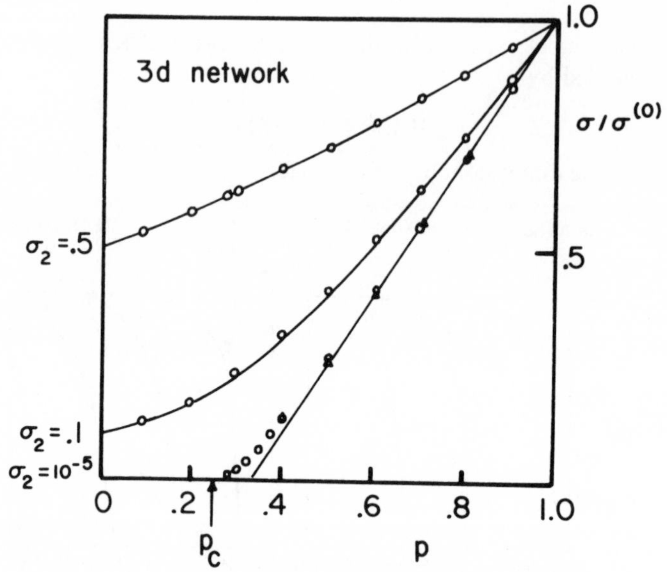

FIGURE 17

Conductivity of a cubic array of conductors with two possible conductances $\sigma_1 = 1$ and $\sigma_2 = .5$ or .1 or 10^{-5} plotted versus the fraction p of bonds with the larger conductance. After Kirkpatrick [**1971**].

from localized states which do not) correlates with the critical probability p_c where p is the fraction of the volume for classically allowed motion. The conductivity is assumed to have the form

$$\sigma(E) = \mu(p(E))P(p(E)), \tag{7.6}$$

where $\mu(p)$ is slowly varying near p_c. Last and Thouless [**1971**] have suggested, however, that the long tortuous paths near $p = p_c$ will cause $\mu(p_c) = 0$. They have supported this conjecture experimentally by measuring the conductance of a piece of conducting paper punched with many small holes. Kirkpatrick [**1972**] has supported the Thouless conclusion by considering two- and three-dimensional networks of resistors containing (at random) two values $\sigma_1 = 1$ and $\sigma_2 = 10^{-5}$. Kirkpatrick finds numerical results in three dimensions of the form [see Figure 17]

$$\sigma = A(p - p_c)^{8/5}, \tag{7.7}$$

whereas, according to (7.5), the exponent should be less than $\frac{2}{3}$. Thus $\mu(p)$ also vanishes at p_c.

TABLE III[1]

Lattice	z	μ	Monte Carlo $p_c(b)$		Monte Carlo $p_c(s)$		Series method $p_c(b)$	Series method $p_c(s)$
Honeycomb	3	1.8484	.640	—	.688	.679	.6527 (exact)	.700
Kagomé	4	—	—	.435	—	.655	—	.6527 (exact)
Square	4	2.6390	.493	.498	.581	.569	.5000 (exact)	.590
Triangular	6	4.1515	.341	.349	.493	.486	.3473 (exact)	.5000 (exact)
Diamond	4	2.878	.390		.436		.388	.425
s.c.	6	4.6826	.254		.325		.247	.307
b.c.c.	8	6.5288	—		—		.178	.243
f.c.c.	12	10.0350	.125		.199		.119	.195
h.c.p.	12	—	.124		.204		—	—

[1] From Shante and Kirkpatrick [**1971**]. z = lattice coordination number.

TABLE IV[2]

Lattice	z	$p_c(b)$	$zp_c(b)$	f	$p_c(s)$	$fp_c(s)$
Honeycomb	3	.6527	1.96	.61	.700	.427
Kagomé	4	—	—	.68	.653	.444
Square	4	.5000	2.00	.79	.590	.466
Triangular	6	.3473	2.08	.91	.500	.455
Diamond	4	.388	1.55	.34	.425	.145
s.c.	6	.247	1.48	.52	.307	.160
b.c.c.	8	.178	1.42	.68	.243	.165
f.c.c.	12	.119	1.43	.74	.195	.144
h.c.p.	12	.124	1.49	.74	.204	.151

[2] From Shante and Krikpatrick [**1971**].

Scher and Zallen [**1970**] have made a contribution to percolation theory in the continuum case by noticing that the highly variable p_c for different discrete lattices (see Table IV) can all be reduced within two percent to a single critical volume v_c where v is the available volume fraction, i.e. the ratio of the volume of the spheres inscribed in the unit cell of the lattice and centered on unblocked sites to that of the crystal. One can equally well define v as

$$v = fp, \tag{7.8}$$

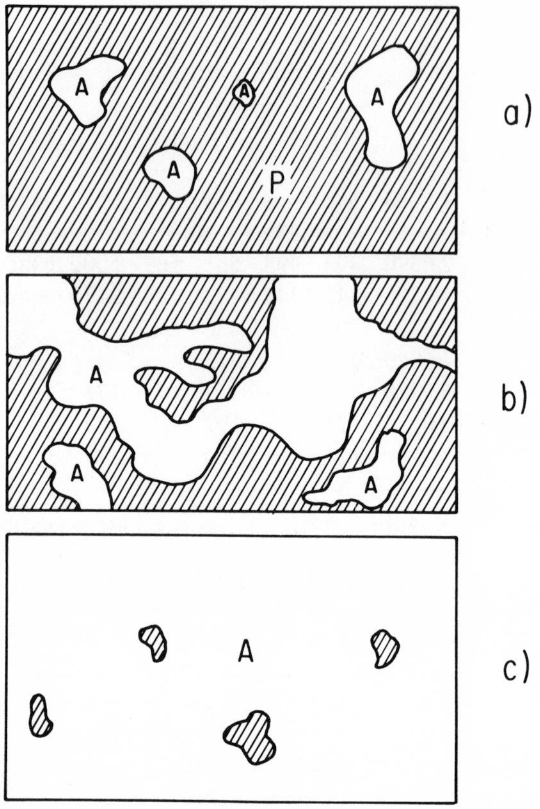

FIGURE 18

The potential $V(r)$ is regarded as the height of a mountain range, with water filled up to a height E. The allowed regions A are those covered with water. At low E, in part a) we have a few lakes and many prohibited (P) continental land regions. This is a condition of localized conductivity. At energy $E = E_c$, in b) the first global body of water appears giving rise to a "mobility edge" at the onset of d.c. conductivity. In c), $E \gg E_c$, most regions are allowed (oceans) with a few islands. This is the extended region d.c. conductivity case.

where f is the filling factor for the lattice, i.e. the ratio of the volume of the inscribed sphere to that of the unit cell and p is the probability that the cell is available, i.e. unblocked. The uniformity of $v_c = fp_c(s)$ from one lattice to another is demonstrated in Table IV. In three dimensions $v_c = .15$ and in two dimensions $v_c = .45$.

D. *Percolation in a continuum.* The continuum percolation problem associated with a classical potential $V(\boldsymbol{r})$ in two dimensions is illustrated in Figure 18. If we visualize $V(\boldsymbol{r})$ as the height of a mountain range, with water filled up to a height E, then the allowed regions A are the regions

$E > V(\mathbf{r})$, i.e. the lakes and/or oceans, whereas the forbidden regions $E < V(\mathbf{r})$ are the islands or continents. In the allowed regions $E > V(\mathbf{r})$ the kinetic energy $E - V(\mathbf{r})$ is positive and classical conduction is possible.

If $P(V)$ is the distribution of potential V [see (9.4) for an evaluation of $P(V)$], then

$$\varphi(E) = \int_{-\infty}^{E} P(V)\, dV \tag{7.9}$$

is the fractional allowed volume and

$$\varphi(E_c) = \varphi_c = .15 \tag{7.10}$$

determines the critical energy (mobility edge, or dividing energy between localized and extended states) in three dimensions. For $E < E_c$ the water level is low enough that the continents are infinite and all allowed (water) areas are lakes. When E is raised to E_c the first ocean appears.

For a model in which V has a Gaussian distribution, the criterion (7.10) yields

$$E_c = \langle V \rangle - 1.03\, V_{rms}. \tag{7.11}$$

In terms of the Halperin-Lax treatment discussed in §4, Zallen and Scher's result is

$$v_c = \frac{\langle V \rangle - E_c}{E_Q} = 1.03 \frac{V_{rms}}{E_Q} = 1.03 \frac{\xi}{E_Q} = 1.03\xi'. \tag{7.12}$$

Contrary to a remark of Zallen and Scher, as long as $\xi' > 1$, the Gaussian statistics of Halperin and Lax are justified and the critical energy v_c is moderately deep in the tail.

E. *Percolation treatment of d.c. hopping conductivity.* Ambegaokar, Halperin and Langer [**1971**] have considered a hopping model of conductivity in an amorphous semiconductor which is represented by a set of conductances G_{ij} connecting the sites of the form

$$G_{ij} = G_0 \exp[-2\alpha R_{ij} - (|E_i| + |E_j| + |E_i - E_j|)/2kT], \tag{7.13}$$

where the separations, R_{ij}, between the sites are random variables as are the energies E_i of each site (relative to the Fermi level).

How should one calculate the conductivity of such a random, three-dimensional network? AHL suggests that the conductivity can be estimated from

$$\sigma = \overline{G}/\overline{R}, \tag{7.14}$$

where $\overline{R}$ is some characteristic length scale for the network, and $\overline{G}$, a characteristic conductance, is taken to be the critical conductance G_c

defined as "the largest value of the conductance such that the subset of resistors with $G_{ij} > G_c$ still contains a connected network which spans the entire system."

The reason supporting the AHL choice is as follows: the low conductance resistors $G_{ij} \ll G_c$ (the dashed lines in Figure 19) make little contribution to the overall d.c. conductivity. If these resistors are removed, there will be a slight reduction in conductivity of the network. The high conductance resistors, $G_{ij} \gg G_c$, the double lines in Figure 19, can be shorted out with only a slight increase in conductivity of the network. Thus the dominant behavior of the conductivity is determined by the critical subnetwork of conductances in the vicinity of G_c.

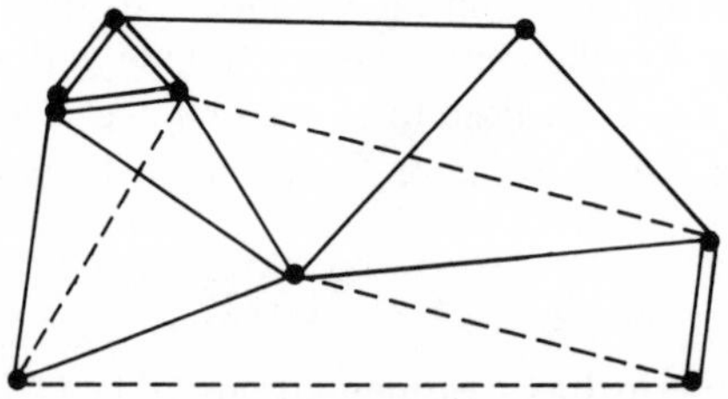

FIGURE 19

A network of small conductances (the dashed lines), of large conductances (the double lines) and a critical subnetwork of intermediate conductances (the solid lines). The dashed lines can be broken in order of increasing conductivity G without changing the conductivity much, as long as an infinite network remains. The critical conductivity G_c is the largest conductivity such that the network with all $G_{ij} > G_c$ is infinitely connected.

AHL find (with reservations as to the applicability of their model) a conductivity,

$$\sigma(T) = A \exp[-(T_0/T)^{1/4}], \tag{7.15}$$

containing the inverse fourth root dependence on temperature which had been predicted by Mott [**1969a**], [**1969c**] and observed in amorphous germanium, silicon and vanadium oxide. (See Ambegaoker, Halperin and Langer [**1971**] for a complete set of references.)

It should not be assumed that such an inverse fourth root dependence follows only from a percolation treatment. Brenig, Döhler and Wölfle [**1971**] have made a variational treatment of the more traditional transport approach and obtained the same result. At low temperatures, the electron succeeds in partly overcoming the Boltzmann factor by hopping longer distances to states closer in energy. This reduces the $1/T$ dependence to a $(1/T)^{1/4}$ dependence.

8. **Master equation and random walk treatment of a.c. and d.c. conductivity.**

A. *Generalized mobility formulas.* Stimulated in 1960 by experimental work of Pollak and Geballe [**1961**], the author started from a rigorous mobility theory (Lax [**1958**]) and a Nyquist formula relating mobility and noise (Lax [**1960**]) and derived an expression for the conductivity σ at frequency ω in terms of the corresponding mobility μ and diffusion constant D:

$$\sigma(\omega) = ne\mu(\omega) = (ne^2/kT)D(\omega) \tag{8.1}$$

where, in the frequency range $\omega \ll kT$, the rigorous diffusion constant is given by

$$D(\omega) = -(\omega^2/6)\int_0^\infty \exp(-i\omega t)\, dt\, \langle[\mathbf{r}(t) - \mathbf{r}(0)]^2\rangle. \tag{8.2}$$

The proof of this formula (Lax [**1961**]) is included in Appendix A of Scher and Lax [**1973**].

At the low concentrations of the Geballe-Pollak [**1961**] experiments, it was permissible to describe the sites $\mathbf{s}$ of the impurity centers by an orthonormalized set of basis functions $\Phi_s = \Phi(\mathbf{r} - \mathbf{s})$. If we let $P(\mathbf{s}, t)$ represent the probability of occupancy of site $\mathbf{s}$ assuming the electron started at the origin, equation (8.2) can be rewritten

$$D(\omega) = -\frac{\omega^2}{6}\sum_{\mathbf{s}} s^2 \int_0^\infty e^{-i\omega t}\, dt\, P(\mathbf{s}, t). \tag{8.3}$$

B. *Transport equation approach.* A typical transport analysis then led to the transport equation

$$\partial P(\mathbf{s}, t)/\partial t = \sum_{\mathbf{s}'} w_{\mathbf{ss}'} P(\mathbf{s}', t) - \left(\sum_{\mathbf{s}'} w_{\mathbf{s}'\mathbf{s}}\right) P(\mathbf{s}, t), \tag{8.4}$$

where the transition rate $w_{\mathbf{ss}'}$ from $\mathbf{s}'$ to $\mathbf{s}$ is a function $w(\mathbf{s} - \mathbf{s}')$ of the impurity separations. This function was calculated carefully by Miller and Abrahams [**1960**] to have the form

$$w_{\mathbf{ss}'} = x^{3/2} \exp(-x) W_0 \Delta[\exp(\Delta/kT) - 1]^{-1}, \tag{8.5}$$

where $r = |\mathbf{s} - \mathbf{s}'|$ and

$$x = 2r/a. \tag{8.6}$$

a is the Bohr radius of the donor state, and $\Delta = \Delta(r, \theta)$ is the energy difference between the final and initial site. If $\Delta > 0$, equation (8.5) describes absorption of a phonon of energy Δ. If $\Delta < 0$, it describes emission of a phonon of energy $|\Delta|$.

Equation (8.3) suggests that we take the one-sided Fourier transform of equation (8.4), which reduces the latter to a set of coupled linear equations (with random coefficients).

C. *The pair approximation.* At the low concentrations of the Pollak-Geballe [**1961**] experiments, the a.c. conductivity is determined by electron hopping between nearest neighbor pairs. In that case, (8.4) reduces to a pair of simultaneous equations. Solution of these equations leads to a diffusion coefficient

$$D(\omega) = \frac{\omega}{12\cosh^2[\Delta/2kT]}\frac{r^2 w(r)}{\omega - iw(r)}, \tag{8.7}$$

where $w(r)$ is the sum of the transition probabilities from atom 1 to 2 and vice-versa:

$$w(r) = Bx^{3/2}\exp(-x) = w(x), \tag{8.8}$$

$$B = W_0\Delta\coth(\Delta/2kT). \tag{8.9}$$

Since typical Δ's are of order 14°K, for the higher temperature data it is permissible to replace Δ by a mean value $\overline{\Delta}$. However, $w(r)$ is very rapidly varying. We should average r over the Hertz nearest neighbor distribution (Appendix 7 of S. Chandrasekhar [**1943**]). If n_D is the density of impurities, the result for the real part of the diffusion constant is

$$\begin{aligned} \operatorname{Re} D(\omega) &= \frac{\omega^2}{12\cosh^2(\overline{\Delta}/2kT)}\left\langle\frac{r^2 w(r)}{\omega^2 + w^2(r)}\right\rangle \\ &= \frac{4\pi n_D(a/2)^5}{12\cosh^2(\overline{\Delta}/2kT)}\,\omega^2\int_0^\infty \frac{x^4\,dx\,w(x)}{\omega^2 + [w(x)]^2}. \end{aligned} \tag{8.10}$$

Since $w(x)$ varies so much more rapidly than x, it is permissible to replace x by a mean value $\bar{x}$ at the maximum of the integral determined by

$$w(\bar{x}) = \omega, \tag{8.11}$$

i.e. the important atomic separations are those for which the jump rate equals the measuring frequency.

Thus

$$\bar{x} \approx \ln(B/\omega) + \tfrac{3}{2}\ln\ln(B/\omega) + \cdots. \tag{8.12}$$

With B of order 10^{12}/sec and ω in the range 10^2–10^6, $\bar{x}$ is a large number. The integral can be approximately evaluated using $w\,dx \approx dw$ with the result

$$\operatorname{Re} D(\omega) \approx \pi^2 n_D(a/2)^5\omega\bar{x}^4/6\cosh^2(\overline{\Delta}/2kT). \tag{8.13}$$

This result was obtained in Lax [**1961**] as are the following quoted comments.

1. $\bar{x}$ is a slowly decreasing function of frequency such that $\omega(\bar{x})^4$ approximates $\omega^{0.8}$ to a surprising accuracy over the experimental range of $\omega/2\pi = 10^2$ to 10^5 cycles.
2. The quantitative results are insensitive to the jump rate constant B, but depend on its general order of magnitude $\sim 10^{12}$/sec determined from equations (2.13) and (2.19) of Miller and Abrahams [**1960**]. It is not surprising then to find agreement on the absolute magnitude within a factor of five (theoretic value being larger).
3. The experimental conductivity in the higher temperature range (where Δ is unimportant) is proportional to $n_A n_D$ in agreement with equations (8.1) and (8.13) using $n = n_A$ since the density of carriers n is equal to the density of acceptors n_A.
4. The factors in the numerator can be given a simple interpretation. The effective jump rate is ω. Hence the diffusion constant should be $\omega\bar{r}^2$ times the probability of finding another donor in the range $(\bar{r}, \bar{r} + a)$, i.e., $4\pi n_D \bar{r}^2 a$, so that $D \sim \omega\bar{r}^4 n_D a \sim \omega\bar{x}^4 n_D a^5$.

Of course, electrons hopping back and forth between two impurities can produce no d.c. current, and $D(\omega)$ does vanish as ω approaches zero. An approximate calculation was made by Scher and Lax [**1969**] in which the pair is surrounded by a continuum with a selfconsistently determined diffusion constant $D(\omega)$. Occasional hops to the continuum permit this model to have an appropriate d.c. as well as a.c. conductivity.

D. *Variational treatments of transport equation.* The kinetic equation (8.4) has been applied by Brenig, Wölfle and Döhler [**1971a**], [**1971b**] to the amorphous semiconductor problem. Using a variational principle, they were able to obtain a d.c. conductivity of the form (7.15) given previously by Mott [**1969**]. Butcher [**1972**] has also made a variational treatment of the kinetic equation, with only preliminary results as yet available.

E. *Continuous time random walk treatment of conductivity.* We have seen that the frequency behavior of the a.c. conductivity is dominated by the large fluctuations in jump rates associated with different nearest neighbor distances. The actual geometrical positions of the impurity sites are relatively unimportant, except for the large distribution of jump times induced by small fluctuations in positions. Let us attempt, therefore, to *simulate* the original problem by placing the impurities on a simple cubic lattice but we must then endow each site with a distribution of hopping times $\psi(t)$ which must be chosen to correspond to the actual distribution of hopping times induced by the distribution of positions of nearby atoms (Scher and Lax [**1972**]).

If $Q(t)$ is the occupancy of a site at the origin and $w(\boldsymbol{j})$ the mean jump rate to a site at $\boldsymbol{j}$, then

$$dQ/dt = -Q \sum_{\boldsymbol{j}} w(\boldsymbol{j}). \tag{8.14}$$

Thus the mean occupancy is given by

$$\langle Q(t)\rangle = \left\langle \exp\left[-\sum_{\boldsymbol{j}} w(\boldsymbol{j})t\right]\right\rangle. \tag{8.15}$$

This requires the evaluation of the characteristic function of a sum of identically distributed random variables. Recent evaluations by Thomas, Hopfield and Augustyniak [**1965**] and by Lax [**1966**], equation (2.30) lead to

$$\langle Q(t)\rangle = \exp\left[-n_D \int d\boldsymbol{r}\,(1 - \exp[-w(\boldsymbol{r})t])\right], \tag{8.16}$$

although this result was undoubtedly known to Markoff according to comments of Chandrasekhar [**1943**] and may have been known to Laplace (see Rice [**1944**]).

If we define

$$\psi(t) = -d\langle Q(t)\rangle/dt, \tag{8.17}$$

then $\psi(t)\,dt$ is the expected rate of transfer out of the site at the origin in the time interval $t, t + dt$. Moreover, since $\langle Q(0)\rangle = 1$,

$$\int_0^\infty \psi(t)\,dt = 1 \tag{8.18}$$

so that $\psi(t)$ is the normalized distribution of hopping times. We may then deduce that

$$\Psi(t) = 1 - \int_0^t \psi(\tau)\,d\tau \tag{8.19}$$

is the probability of no jump in the time interval $(0, t)$. The probability that the nth hop will occur in $(t, t + dt)$, namely, $\psi_n(t)\,dt$ can be calculated by iteration

$$\psi_n(t) = \int_0^t \psi(\tau)\psi_{n-1}(t - \tau)\,d\tau. \tag{8.20}$$

If $P_n(\boldsymbol{s})$ is the probability of being found at site $\boldsymbol{s}$ after the nth hop, then the usual *discrete* random walk description leads to a recursion relation of the form

$$P_{n+1}(\boldsymbol{s}) = \sum_{\boldsymbol{s}'} p(\boldsymbol{s} - \boldsymbol{s}')P_n(\boldsymbol{s}'), \tag{8.21}$$

where $p(s - s')$ is the probability of jumping from site s' to site s. The continuous time nature of the problem can be introduced following the technique of Montroll and Weiss [**1965**] and Montroll ([**1964**], [**1967**], [**1969**]) by defining $Q(s, \tau)\,d\tau$ to be the probability of reaching s in the interval $[\tau, \tau + d\tau]$. Adding up the contributions to reaching s over all possible numbers, n, of steps, Montroll and Weiss obtain

$$Q(s, \tau)\,d\tau = \sum_n P_n(s)\psi_n(\tau)\,d\tau. \tag{8.22}$$

Combining this result with the probability of remaining at s from τ to t we get

$$P(s, t) = \int_0^t Q(s, \tau)\Psi(t - \tau)\,d\tau = \sum_n P_n(s) \int_0^t \psi_n(\tau)\Psi(t - \tau)\,d\tau. \tag{8.23}$$

(Further discussion of the relation between master equations and random walks is given by Bedeaux et al. [**1971**]).

Since $\psi_n(t)$ is, by (8.19), an n-fold convolution of $\psi(t)$, it is easy to take the Laplace transform of equation (8.23). But equation (8.3) tells us that it is the second moment of the Laplace transform that is needed. Moreover, the random walk problem for $P_n(s)$ can be solved exactly. A simple closed form expression is thus found (Scher and Lax [**1972**], [**1973**]), for the diffusion constant (8.3),

$$D(\omega) = \frac{\sigma^2}{6} \frac{i\omega\phi(\omega)}{1 - \phi(\omega)}, \tag{8.24}$$

where

$$\phi(\omega) = \int_0^\infty e^{-i\omega t}\psi(t)\,dt \tag{8.25}$$

is the one-sided Fourier transform of the hopping distribution, and

$$\sigma^2 = \sum_s s^2 p(s) \tag{8.26}$$

is the second spatial moment of the probability $p(s)$ of a hop s in a single step.

Thus we have a closed form solution of the continuous time random walk problem. The implementation of this solution is nontrivial, since we must obtain the Laplace transform (8.25) of $\psi(t)$ which in turn requires the evaluation of (8.15) using the expression (8.5) for $w(r)$. Details are presented in Scher and Lax ([**1972**], [**1973**]).

As expected from the pair approximation, $D(\omega)$ is found to be a very smooth function of ω (varying slower than ω) for many decades. The

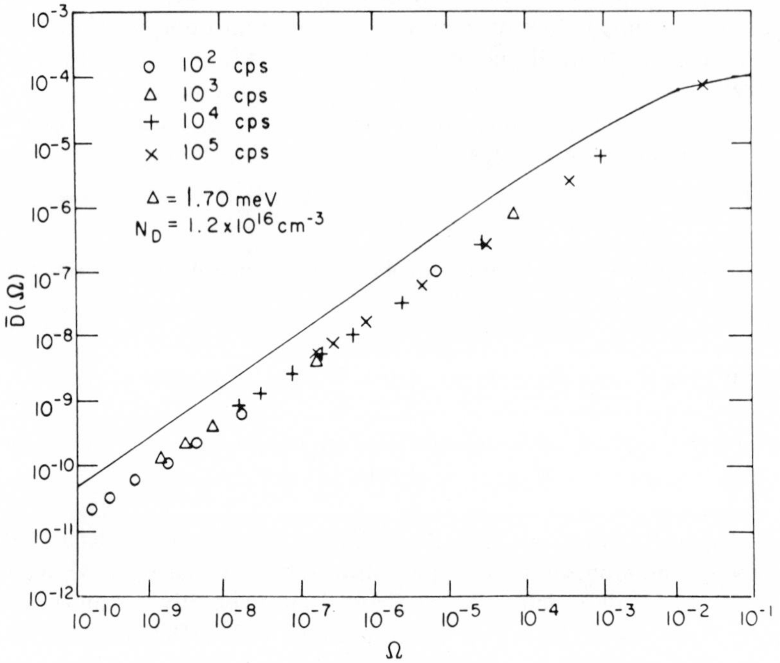

FIGURE 20

A plot against dimensionless frequency $\Omega = \omega/[W_M e^{\gamma}]$ of the real part of the experimental and theoretical dimensionless diffusion constant

$$\bar{D}(\Omega) = 6D(\omega)/(\sigma^2 W_M),$$

where $\gamma = .5772$, σ^2 is the mean-square hopping distance, (8.24), and W_M is a mean value for the maximum hopping rate such that equation (8.6) can be rewritten as $w(\mathbf{r}) = W_M \exp(-2r/a)$ where a is the Bohr radius of the impurity center. Experimental results were carried out at a donor concentration $n_D = 1.2 \times 10^{-16}\,\text{cm}^{-3}$, with an experimental activation energy $\Delta = 1.7$ meV over the low temperature range $10/T > 1.7$.

experimental data were found to have two activation energies. Thus low and high temperature data are plotted separately in Figures 21 and 22 and compared with our numerical calculations of $D(\omega)$. Excellent agreement has been obtained.

To obtain the d.c. conductivity, it is necessary to take the limit as ω approaches zero of equation (8.24). For small ω, we note that

$$\phi(\omega) \to 1 - i\omega\bar{t}, \tag{8.27}$$

where

$$\bar{t} = \int_0^\infty t\psi(t)\,dt = \int_0^\infty \langle Q(t)\rangle\,dt \tag{8.28}$$

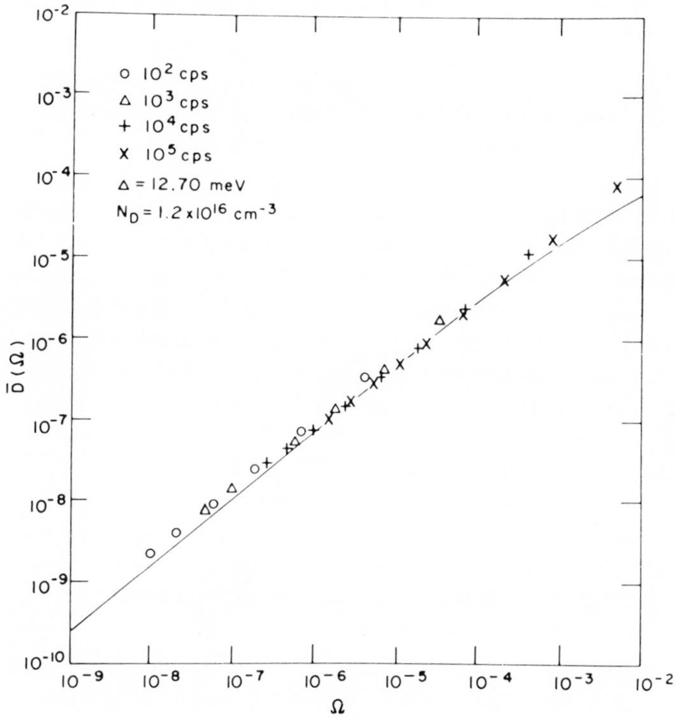

FIGURE 21

Comparison of experiment and theory of Re $\bar{D}(\Omega)$ vs. Ω in the high temperature range $10/T < 1.14$ for which the activation energy is found to be $\Delta = 12.7$ meV. See caption to Figure 20.

is the mean time between hops. In summary, a closed form expression is obtained for the d.c. diffusion constant

$$D(0) = \sigma^2/(6\bar{t}), \tag{8.29}$$

where

$$\bar{t} = \int_0^\infty dt \left\langle \exp\left[-n_D \int d\boldsymbol{r}(1 - \exp[-w(\boldsymbol{r})t])\right]\right\rangle \tag{8.30}$$

and $w(\boldsymbol{r})$ is the hopping rate defined by equation (8.5). The average indicated in (8.28) is over the distribution of energies Δ.

The above analysis has recently been generalized (Scher and Lax [**1973**]) to correlate space and time at each hop by using a more general distribution $\psi(\boldsymbol{s}, t)$ to represent the probability of a hop of displacement $\boldsymbol{s}$ in the interval $(t, t + dt)$.

A detailed comparison of equation (8.28) with experiment is made in Scher and Lax [**1973**].

9. **Other methods.** Since we cannot make detailed comments on all the methods used, we should like to call the readers' attention here to additional methods of importance in random systems.

A. *Moment methods.* Since it is often possible to calculate exactly the first few moments of the density of states, a number of authors have attempted to fit the density of states using a suitable functional form with a few parameters. Montroll ([**1942**], [**1943**]) was an early advocate of this method for lattice vibrations in perfect crystals. Lax and Lebowitz [**1954**] showed that this method could be highly successful if functions containing the correct singularities were introduced. Kane [**1962**] related the moments to perturbation theory and applied his results (Kane [**1963a**]) to the Lax-Phillips model of an impurity band. He wisely did not attempt to use his results at low density since there the density of states has a singularity as shown in Table I. He also did not attempt a fit at high densities since his perturbation expansion would not give him a good estimate of the moments. At the intermediate density $\varepsilon = 1$, a qualitative fit was obtained, but not as good as the coherent potential approximation.

In spite of the difficulty of fitting a function by just a few moments, this method will be popular because it is often the only *simple* way to estimate the density of states. See, for example, Brinkman and Rice [**1970**].

B. *The smoothing method.* The smoothing method (Frisch [**1968**], Bourret [**1962**]–[**1968**]) has been given an especially simple rigorous description by Keller [**1964**]. The random part of the perturbation $V(\boldsymbol{r}) - \langle V(\boldsymbol{r})\rangle$ is treated as small. The incoherent part of the wave function $\psi - \langle\psi\rangle$ is then eliminated in first order to obtain an equation for $\langle\psi\rangle$ correct to second order in the potential.

McKenna and Morrison [**1971**] have shown, however, by exactly solving the problem for a sum of two (or more) telegraph waves, that the smoothing method is only valid in the limit of weak random perturbations.

Keller [**1964**] has reconsidered the problem as a multiple scattering problem, and has shown that when the transition matrix (not necessarily the perturbation) is weak, he obtains a result in agreement with the variational approximation, (B23), to the Lax [**1952**] quasicrystalline approximation. The latter approximation is, of course, not restricted to weak t matrices since it is exact for a perfect crystal. This QCA method might be expected to be poorest for a random distribution of strong scatterers. In this case, however, we have shown in Appendix B that it reduces to the

CPA and is in moderately good agreement with the exact Lax-Phillips results in Figure 5, except deep in the tail.

C. *Semiclassical approximation.* Kane [**1963b**], Bonch-Bruevich [**1962**], Lifshitz [**1968**] and Eggarter [**1972**] have made semiclassical or Fermi-Thomas type approximations to the density of states. In this treatment, the potential energy is regarded as varying sufficiently slowly that a "local density of states can be defined just as if the potential were constant." For a constant V the density of states per unit volume in three dimensions can be written

$$\rho(E, V) = \frac{1}{(2\pi)^3}\int d\boldsymbol{k}\,\delta(E - V - \hbar^2\boldsymbol{k}^2/2m) = \frac{m^{3/2}}{2^{1/2}\hbar^3}(E - V)^{1/2}. \tag{9.1}$$

(An extra factor of 2 should be added to account for up and down spin states.) The density of states can then be written

$$\rho(E) = \int \rho(E, V)P(V)\,dV, \tag{9.2}$$

where $P(V)$ is the probability density of the total potential V. If

$$V = \sum_j v(\boldsymbol{r} - \boldsymbol{r}_j), \tag{9.3}$$

the distribution of V can be obtained by the method of characteristic functions. The latter can be evaluated by the method used in going from (8.15) to (8.16). The result, as given by Rice ([**1943**], [**1944**]) is

$$P(V) = \frac{1}{2\pi}\int_{-\infty}^{\infty} d\alpha\,\exp\left\{i\alpha V + n\int [e^{-i\alpha v(\boldsymbol{r})} - 1]\,d\boldsymbol{r}\right\}, \tag{9.4}$$

where n is the average concentration of centers.

Equation (9.4) is descriptive of Poisson statistics, but at high concentrations, a Gaussian is a good approximation. This leads in turn to a Gaussian tail to the main part of the density of states. The more general behavior $\exp(-|E|^n)$ with n varying from $\frac{1}{2}$ to 2 found in §4 is due to the fact that the minimum counting method there took account of the increase in kinetic energy associated with localization in the tail states.

Acknowledgments. The author wishes to acknowledge computational help from Mrs. B. C. Chambers, and a careful reading of the manuscript by B. I. Halperin and E. O. Kane.

Appendix A. Derivation of multiple scattering equations. Equation (3.11) can be rewritten in the form

$$(E - H_0)\Psi = \sum_j v_j\Psi. \tag{A1}$$

If there is an external incident wave Φ, which is a solution of the unperturbed equation $(E - H_0)\Phi = 0$, the solution to (A1) can be written

$$\Psi = \Phi + (E - H_0)^{-1} \sum_j v_j \Psi = \Phi + \sum_j L_j, \tag{A2}$$

where L_j is the wave which leaves site j. Thus the exciting or effective field Ψ^i can be defined as

$$\Psi^i = \Phi + \sum_{j \neq i} L_j, \tag{A3}$$

so that $\Psi = \Psi^i + L_i$ can be rewritten as

$$\Psi - (E - H_0)^{-1} v_i \Psi = \Psi^i. \tag{A4}$$

Let us seek a solution of (A4) of the form

$$\Psi = (1 + R_i)\Psi^i = \Psi^i + R_i\Psi^i = \Psi^i + L_i. \tag{A5}$$

Equation (A4) then yields the condition

$$R_i - (E - H_0)^{-1} v_i R_i = (E - H_0)^{-1} v_i \tag{A6}$$

which suggests setting

$$R_i = (E - H_0)^{-1} \cdot t_i \tag{A7}$$

resulting in the expected equation

$$t_i = v_i + v_i (E - H_0)^{-1} t_i \tag{A8}$$

for the transition operator. Equation (A5) now yields

$$L_i = R_i \Psi^i = (E - H_0)^{-1} t_i \Psi^i. \tag{A9}$$

This result substituted into (A2) and (A3) yields our working equations (3.12) and (3.13).

Our renormalized propagator formulation, equations (3.23)–(3.25), can be written as

$$(E - H - V)G = 1, \quad \text{or} \quad (E - H)G = 1 + \sum_j V_j G, \tag{A10}$$

or finally

$$G = G_c + (E - H)^{-1} \sum_j V_j G, \tag{A11}$$

where $G_c = (E - H)^{-1}$. Equation (A11) is exactly parallel to (A2) and the procedure which led to (A8) and (A9) yields

$$T_i = V_i + V_i (E - H)^{-1} T_i, \tag{A12}$$

$$L_j = (E - H)^{-1} T_j G^j, \tag{A13}$$

and equations (3.27) and (3.28) in the text.

Appendix B. Coherent potentials in the quasicrystalline approximation. The influence of correlations on the coherent potential was previously obtained in an unmodified propagator formalism. The results obtained (equations (3.18) and (4.2) of Lax [**1952**]) can be immediately rewritten in the modified propagator formalism merely by replacing H_0 by H, v_j by V_j, t_j by $\hat{t}_j$, etc. Since these results were obtained using an infinite medium with no external sources, we shall rederive them, here, directly from the Green's function equations (3.27) and (3.28) of the modified propagator formalism which has G_c as a driving term.

The ensemble average of (3.27),

(B1) $$\langle G \rangle = G_c + G_c \sum_j \langle T_j G^j \rangle = G_c + G_c \int n(\boldsymbol{j})\, d\boldsymbol{j}\, T_j \langle G^j \rangle_j,$$

has been performed by first averaging over all positions but that of the jth scatterer (which is the meaning of the symbol $\langle\ \rangle_j$) followed by an average over the jth scatterer. In a liquid, $n(\boldsymbol{j}) = n =$ a constant. The relationship (3.10) can be written in operator form

(B2) $$T_j = \exp(-i\boldsymbol{p} \cdot \boldsymbol{j}) T \exp(i\boldsymbol{p} \cdot \boldsymbol{j})$$

where $\boldsymbol{p} = -i\nabla$ is the momentum operator of quantum mechanics.

Equation (B1) is an operator equation, which we have seen in §3 is diagonal in the $\boldsymbol{k}$ representation. If we let $\langle\ |\boldsymbol{k}\rangle$ be the Dirac ket vector for a plane wave, i.e.

(B3) $$\langle \boldsymbol{r} | \boldsymbol{k} \rangle = \exp(i\boldsymbol{k} \cdot \boldsymbol{r}),$$

where we have normalized in a box of unit volume, then we can set

(B4) $$\langle \boldsymbol{r} | G_c | \boldsymbol{k} \rangle = (E - \tfrac{1}{2}k^2 - V_{ckk})^{-1} \exp(i\boldsymbol{k} \cdot \boldsymbol{r}),$$

(B5) $$\langle \boldsymbol{r} | G^j | \boldsymbol{k} \rangle = G^j(\boldsymbol{r}, \boldsymbol{j}),$$

(B6) $$\exp(i\boldsymbol{p} \cdot \boldsymbol{j}) G^j(\boldsymbol{r}, \boldsymbol{j}) = G^j(\boldsymbol{r} + \boldsymbol{j}, \boldsymbol{j}) = \exp(i\boldsymbol{k} \cdot \boldsymbol{j}) g^e(\boldsymbol{r}),$$

where the last step follows from invariance of the ensemble average to displacements as shown in equation (3.3) of Lax [**1952**]. In this mixed representation, the second term of (B1) can be written in the form

(B7) $$D T(0) g^e(\boldsymbol{r}),$$

where

(B8) $$D = n \int \exp[i(\boldsymbol{k} - \boldsymbol{p}) \cdot \boldsymbol{j}]\, d\boldsymbol{j}$$

has only diagonal matrix elements

(B9) $$D_{ba} = n\delta(\boldsymbol{b}, \boldsymbol{a})\delta(\boldsymbol{k}, \boldsymbol{a}),$$

thus insuring that $\langle G \rangle$ is diagonal as claimed.

The coherent potential V_c is determined by the requirement that

(B10) $$\langle G \rangle = G_c.$$

Since these operators are diagonal, this merely requires that the $\boldsymbol{k}, \boldsymbol{k}$ matrix element of the second term in (B1) vanish. In view of (B7)–(B9) together with (3.36) we obtain the result

(B11) $$\langle \boldsymbol{k}|V_c|\boldsymbol{k}\rangle = n\frac{\langle \boldsymbol{k}|\hat{t}g^e(\boldsymbol{r})\rangle}{\langle \boldsymbol{k}|g^e(\boldsymbol{r})\rangle} = n\frac{(\exp(i\boldsymbol{k}\cdot\boldsymbol{r}), \hat{t}g^e(\boldsymbol{r}))}{(\exp(i\boldsymbol{k}\cdot\boldsymbol{r}), g^e(\boldsymbol{r}))}$$

previously given in equation (3.18) of Lax [**1952**], with a minor change of notation.

The ensemble average of (3.28) yields

(B12) $$\langle G^j \rangle_j = G_c + G_c \sum_{i \neq j} \langle T_i G^i \rangle_{ij} = G_c + G_c \int n(\boldsymbol{j}|\boldsymbol{i})\, d\boldsymbol{i}\; T_i \langle G^i \rangle_{ij},$$

which $\langle\ \rangle_{ij}$ implies an average over all but particles i and j, and $n(\boldsymbol{j}|\boldsymbol{i})$ is the density at $\boldsymbol{i}$ knowing there is a particle at $\boldsymbol{j}$. In a crystal, the extra information that j is fixed as well as i does not change the situation in which all particles are fixed anyway. Thus the quasicrystalline approximation takes the form

(B13) $$\langle G^i \rangle_{ij} \approx \langle G^i \rangle_i.$$

Once this approximation is made, it is necessary to take account of correlations. If $n(\boldsymbol{j}|\boldsymbol{i})$ were to be replaced by n, the second term in (B12) would vanish, as this was the condition previously used to determine V_c. Thus the improvement of the QCA over the CPA is only evident when correlations are included. The $\langle \boldsymbol{r}|\ |\boldsymbol{k}\rangle$ matrix element of (B12) yields

(B14) $$G^j(\boldsymbol{r}, \boldsymbol{j}) = G_c(\boldsymbol{r}) + \int d\boldsymbol{i}\,[n(\boldsymbol{j}|\boldsymbol{i}) - n]e^{-i\boldsymbol{p}\cdot\boldsymbol{i}}TG^i(\boldsymbol{r} + \boldsymbol{i}, \boldsymbol{i}),$$

where the term in n, which yields a vanishing contribution, has been subtracted off to improve convergence of the integral. Replacing $\boldsymbol{r}$ by $\boldsymbol{r} + \boldsymbol{j}$, we obtain the integral equation

(B15) $$g^e(\boldsymbol{r}) = G_c(\boldsymbol{r}) + G_c M(\boldsymbol{p} - \boldsymbol{k})Tg^e(\boldsymbol{r}),$$

where

(B16) $$M(\boldsymbol{p} - \boldsymbol{k}) = \int [n(0|\boldsymbol{r}') - n]\, d\boldsymbol{r}' \exp[i(\boldsymbol{k} - \boldsymbol{p})\cdot\boldsymbol{r}'],$$

where $\boldsymbol{r}' = \boldsymbol{i} - \boldsymbol{j}$. Equations (B15) and (B16) are identical to equations (4.5) and (4.6) of Lax [**1952**] except for a shift from bare to modified propagators.

Equation (B11) can be replaced by the simpler equation

(B17) $$V_{ckk} = n(\exp(i\boldsymbol{k}\cdot\boldsymbol{r}), \hat{t}\psi_{\boldsymbol{k}}^{e}(\boldsymbol{r})),$$

where the normalization denominator in (B11) has been cancelled by defining ψ^e to be the solution of

(B18) $$\psi_{\boldsymbol{k}}^{e}(\boldsymbol{r}) = \exp(i\boldsymbol{k}\cdot\boldsymbol{r}) + L\hat{t}\psi_{\boldsymbol{k}}^{e}(\boldsymbol{r}),$$

(B19) $$L = (E - H)^{-1}M(\boldsymbol{p} - \boldsymbol{k}).$$

If the solution of the adjoint equation obeys

(B20) $$\psi_{\boldsymbol{a}}^{e}(\boldsymbol{r})' = \exp(i\boldsymbol{a}\cdot\boldsymbol{r}) + L^{\dagger}(\hat{t})^{\dagger}\psi_{\boldsymbol{a}}^{e}(\boldsymbol{r})',$$

then according to Lax [**1950**] a variational expression for the coherent potential is given by

(B21) $$V_{ckk} = \frac{n(\psi_{\boldsymbol{k}}^{e}(\boldsymbol{r})', \hat{t}\exp(i\boldsymbol{k}\cdot\boldsymbol{r}))(\exp(i\boldsymbol{k}\cdot\boldsymbol{r}), \hat{t}\psi_{\boldsymbol{k}}^{e}(\boldsymbol{r}))}{(\psi_{\boldsymbol{k}}^{e}(\boldsymbol{r})', (\hat{t} - \hat{t}L\hat{t})\psi_{\boldsymbol{k}}^{e}(\boldsymbol{r}))}.$$

This equation is the analogue of equation (4.7) of Lax [**1952**], a variational principle in the corresponding bare propagator case.

We can always define an effective field correction factor c by

(B22) $$V_{ckk} = n\hat{t}_{kk}c.$$

If we approximate $\psi_{\boldsymbol{k}}^{e}(\boldsymbol{r}) = \psi_{\boldsymbol{k}}^{e}(\boldsymbol{r})' = \exp(i\boldsymbol{k}\cdot\boldsymbol{r})$ we get

(B23) $$c = (1 - J)^{-1},$$

where

(B24) $$J = \frac{1}{\hat{t}_{kk}}\int \frac{d\boldsymbol{b}}{(2\pi)^3}\frac{M(\boldsymbol{b} - \boldsymbol{k})\hat{t}_{kb}\hat{t}_{bk}}{E - \hbar^2\boldsymbol{b}^2/2m - V_{cbb}}.$$

In contrast with equation (4.10) of Lax [**1952**], equations (B22)–(B24) constitute a selfconsistent determination of the coherent potential V_{ckk}.

Appendix C. Solution of the single scattering integral equation. For a single scatterer at the origin describable by an attractive delta function potential

(C1) $$v(x) = -|V_0|\delta(x).$$

Equation (3.39) can be rewritten in matrix form as

(C2) $\langle x|\hat{t}|x'\rangle = -|V_0|\delta(x)\langle x|x'\rangle - |V_0|\delta(x)\int \langle 0|e^{-1}|x''\rangle\, dx''\langle x''|\hat{t}|x'\rangle.$

Since the right-hand side contains $\delta(x)$ as a common factor, we can assume a solution of the form

(C3) $$\langle x|\hat{t}|x'\rangle = \delta(x)F(x').$$

The integral equation (C2) then reduces to an algebraic equation whose solution is

(C4) $$F(x') = -|V_0|\langle 0|x'\rangle/[1 + |V_0|\langle 0|e^{-1}|0\rangle].$$

If we note that

(C5) $$-|V_0|\delta(x)\langle 0|x'\rangle = \langle x| - |V_0|\delta(x)|x'\rangle = \langle x|v(x)|x'\rangle,$$

equation (C3) can be rewritten in operator form

(C6) $$\hat{t} = v(x)/[1 + |V_0|\langle 0|e^{-1}|0\rangle].$$

To evaluate the matrix element in the denominator we note that

(C7) $$\langle x|e^{-1}|x'\rangle = \frac{1}{2\pi}\int \frac{\exp[ik(x - x')]\, dx}{E - V_c - (\hbar^2/2m)k^2} = -\frac{2m}{\hbar^2}\frac{\exp[-K'|x - x'|]}{2K'},$$

where

(C8) $$(K')^2 = (2m/\hbar^2)(V_c - E).$$

With the definition,

(C9) $$K_0 = m|V_0|/\hbar^2,$$

(C10) $$\hat{t} = v(x)/[1 - (K_0/K')].$$

If we set $\hbar = m = K_0 = 1$, equation (C10) reduces to (3.51).

References

V. Ambegaokar, B. I. Halperin and J. S. Langer
1971. Phys. Rev. **B4** (1971), 2612.

P. W. Anderson
1958. Phys. Rev. **109** (1958), 1492.
1970. *Comments on solid state physics* **2** (1970), 193.

I. G. Austin and N. F. Mott
1969. Advances in Physics **18** (1969), 41.

D. Bedeaux, K. Lakatos-Lindenberg and K. E. Shuler

J. L. Beeby
1964. Proc. Phys. Soc. (London) **A279** (1964), 82; Phys. Rev. **135** (1964), A130.

V. L. Bonch-Bruevich
1962. Proc. Internat. Conference on the Physics of Semiconductors (Exeter, July 1962),

Institute of Physics and the Physical Society, London, 1962, p. 216.
1966. *Physics of* III-V *Compounds*. Vol. 1, Academic Press, New York, 1966, Chap. 4.

R. C. Bourret
1962. *Stochastically perturbed fields, with application to wave propagation in random media*, Nuovo Cimento (10) **26** (1962), 1–31. MR **26** #2276.
1962. *Propagation of randomly perturbed fields*, Canad. J. Phys. **40** (1962), 782–790. MR **27** #6473.
1965. *Quantised fields as random classical fields*, Phys. Lett. **12** (1964), 323–325. MR **32** #814.
1965. *Ficton theory of dynamical systems with noisy parameters*, Canad. J. Phys. **43** (1965), 619–639. MR **31** #1062.
1966. Canad. J. Phys. **44** (1966), 2519.

W. Brenig, G. Döhler and P. Wölfle
1971. Phys. Lett. **35A** (1971), 77; Z. Physik **246** (1971), 1.

W. F. Brinkman and T. M. Rice
1970. Phys. Rev. **B2** (1970), 1324.

F. Brouers
1970. J. Non-Cryst. Solids **4** (1970), 428.

P. N. Butcher
1972. Private communication.

S. Chandrasekhar
1943. *Stochastic problems in physics and astronomy*, Rev. Modern Phys. **15** (1943), 1–89. MR **4,** 248.

M. H. Cohen, H. Fritsche and S. R. Ovshinsky
1969. Phys. Rev. Lett. **22** (1969), 1065.

M. Cutler and N. F. Mott
1969. Phys. Rev. **181** (1969), 1336.

P. Dean
1961. Proc. Roy. Soc. London Ser. A **260** (1961), 263.

P. Dean and M. D. Bacon
1965. *Vibrations of two-dimensional disordered lattices*, Proc. Roy. Soc. Ser. A **283** (1965), 64–82. MR **31** #1079.

E. N. Economou and M. H. Cohen
1972. Phys. Rev. **B5** (1972), 1931.

E. N. Economou, M. H. Cohen, K. F. Freed and E. S. Kirkpatrick
1973. Chapter in *Amorphous and liquid semiconductors* (J. Tauc, editor), Plenum Press, New York, 1973.

E. N. Economou, S. Kirkpatrick, M. H. Cohen and T. P. Eggarter
1970. Phys. Rev. Lett. **25** (1970), 520.

S. Edwards
1970. J. Phys. **C3** (1970), L30; J. Non-Cryst. Solids **4** (1970), 417.

T. P. Eggarter
1972. Phys. Rev. **5A** (1972), 2496.

T. P. Eggarter and M. H. Cohen
1970. Phys. Rev. Lett. **25** (1970), 807.
1971. Phys. Rev. Lett. **27** (1971), 129.

E. Feenberg
1948. *A note on perturbation theory*, Phys. Rev. (2) **74** (1948), 206–208. MR **10,** 154.

H. Feshbach
1958. Ann. Rev. Nuclear Sci. **8** (1958), 49; Ann. Physics **5** (1958), 357.

R. P. Feynman
1948. *Space-time approach to non-relativistic quantum mechanics*, Rev. Modern Physics **20** (1948), 367–387. MR **10,** 224.
R. P. Feynman and A. R. Hibbs
1965. *Quantum mechanics and path integrals*, McGraw-Hill, New York, 1965.
L. L. Foldy
1945. *The multiple scattering of waves.* I. *General theory of isotropic scattering by randomly distributed scatterers*, Phys. Rev. (2) **67** (1945), 107–119. MR **6,** 224.
H. L. Frisch, J. M. Hammersley and D. J. A. Welsh
1962. Phys. Rev. **126** (1962), 949.
H. L. Frisch and J. M. Hammersley
1963. *Percolation-processes and related topics*, J. Soc. Indust. Appl. Math. **11** (1963), 894. MR **28** #2584.
H. L. Frisch and S. P. Lloyd
1960. Phys. Rev. **120** (1960), 1175.
H. L. Frisch, E. Sonnenblick, V. A. Vyssotsky and J. M. Hammersley
1961. Phys. Rev. **124** (1961), 1021.
U. Frisch
1968. *Probabilistic methods in applied mathematics*, A. T. Bharucha-Reid (editor), Academic Press, New York, 1968.
I. M. Gel'fand and A. M. Jaglom
1960. *Integration in functional spaces and its applications in quantum physics*, Uspehi Mat. Nauk **11** (1956), no. 1 (67), 77–114; English transl., J. Mathematical Phys. **1** (1960), 48–69. MR **17,** 1261; MR **22** #3455.
M. L. Goldberger and K. M. Watson
1964. *Collision theory*, Wiley, New York, 1964. MR **29** #3128.
B. L. Gyorffy
1970. Phys. Rev. **B1** (1970), 3290.
1972. Phys. Rev. **B5** (1972), 2382.
B. I. Halperin
1966. Phys. Rev. **139** (1966), A104.
1971. *Lecture notes for Summer School on the theory of condensed matter*, Kiljava, Finland, August 1971.
B. I. Halperin and M. Lax
1966. Phys. Rev. **148** (1966), 722.
1967. Phys. Rev. **153** (1967), 802.
J. H. Hammersley
1961. *Comparison of atom and bond percolation processes*, J. Mathematical Phys. **2** (1961), 728–733. MR **24** #A582.
R. Z. Has'minskiĭ
1966. *A limit theorem for solutions of differential equations with a random right hand part*, Teor. Verojatnost. i Primenen. **11** (1966), 444–462 = Theor. Probability Appl. **11** (1966), 390–406. MR **34** #3637.
E. O. Kane
1962. *Spectral moments and continuum perturbation theory*, Phys. Rev. (2) **125** (1962), 1094–1099. MR **25** #949.
1963a. Phys. Rev. **131** (1963), 1532.
1963b. Phys. Rev. **131** (1963), 79.
J. R. Klauder
1961. Ann. Phys. **14** (1961), 43.

J. B. Keller
1962. *Wave propagation in random media*, Proc. Sympos. Appl. Math., vol. 13, Amer. Math. Soc., Providence, R.I., 1962, pp. 227–246. MR **25** #3683.
1964. *Stochastic equations and wave propagation in random media*, Proc. Sympos. Appl. Math., vol. 16, Amer. Math. Soc., Providence, R.I., 1964, pp. 145–170. MR **31** #2895.

S. Kirkpatrick
1971. Phys. Rev. Lett. **27** (1971), 1722.

W. Kohn and N. Rostoker
1954. Phys. Rev. **94** (1954), 1111.

J. Korringa
1947. *On the calculation of the energy of a Bloch wave in a metal*, Physica **13** (1947), 392–400. MR **9,** 401.

B. J. Last and D. J. Thouless
1971. Phys. Rev. Lett. **27** (1971), 1719.

M. Lax
1950. Phys. Rev. **78** (1950), 306.
1951. *Multiple scattering of waves*, Rev. Modern Physics **23** (1951), 287–310. MR **13,** 708.
1952. Phys. Rev. **85** (1952), 621.
1958. *Generalized mobility theory*, Phys. Rev. (2) **109** (1958), 1921–1926. MR **19,** 1207.
1960. Rev. Modern Phys. **32** (1960), 25.
1961. Typed unpublished manuscript, Presented at Jan. 1962 *Conference on "Electrons in Disordered Structures"* at Cambridge, England.
1966. Rev. Modern Phys. **38** (1966), 541.

M. Lax and J. Lebowitz
1954. Phys. Rev. **96** (1954), 594.

M. Lax and J. C. Phillips
1958. *One-dimensional impurity bands*, Phys. Rev. (2) **110** (1958), 41–49. MR **23** #B1986.

P. Leath
1972. Phys. Rev. **B5** (1972), 1643.

I. M. Lifshitz
1965. *Energy spectrum structure and quantum states of disordered condensed systems*, Uspehi Fiz. Nauk **83** (1964), 617–663 = Soviet Physics Uspehi **7** (1965), 549–573. MR **31** #5597.
1968. Soviet Physics JETP **26** (1968), 462.

P. Lloyd
1957. J. Phys. **C2** (1957), 1717.

T. Matsubara and Y. Toyozawa
1961. *Theory of impurity band conduction in semiconductors. An approach to random lattice problem*, Progr. Theoret. Phys. **26** (1961), 739–756. MR **24** #B2434.

J. McKenna and J. A. Morrison
1971. J. Mathematical Phys. **12** (1971), 2126.

H. Miller and E. Abrahams
1960. Phys. Rev. **120** (1960), 745.

E. W. Montroll
1964. *Random walks on lattices*, Proc. Sympos. Appl. Math., vol. 16, Amer. Math. Soc., Providence. R.I., 1964, pp. 193–220. MR **28** #4585.
1967. in *Energetics in metallurgical phenomena*. Vol. 3, Gordon and Breach, New York, 1967, p. 123.
1969. J. Mathematical Phys. **10** (1969), 753.

E. W. Montroll and G. H. Weiss

1965. *Random walks on lattices.* II, J. Mathematical Phys. **6** (1965), 167–181. MR **30** #2563.

J. Morrison

1962. J. Mathematical Phys. **3** (1962), 1.

1972. J. Math. Anal. Appl. **38** (1972), 13.

P. M. Morse and H. Feshbach

1953. *Methods of theoretical physics*, McGraw-Hill, New York, 1953. MR **15,** 583.

N. F. Mott

1966. Philos. Mag. **13** (1966), 989.

1967. Adv. Phys. **16** (1967), 49.

1968a. Philos. Mag. **17** (1968), 1259.

1968b. Rev. Modern Phys. **40** (1968), 677.

1968c. J. Non-Cryst. Solids **1** (1968), 1.

1969a. Philos. Mag. **19** (1969), 835.

1969b. Contemp. Phys. **10** (1969), 125.

1969c. Festkorperprobleme **9** (1969), 22.

1970. Philos. Mag. **22** (1970), 7.

N. F. Mott and W. D. Twose

1961. Adv. Phys. **10** (1961), 107.

N. F. Mott and R. S. Allgaier

1967. Phys. Stat. Sol. **21** (1967), 343.

N. F. Mott and E. A. Davis

1968. Philos. Mag. **17** (1968), 1269.

B. G. Nickel and J. A. Krumhansl

1971. Phys. Rev. **B4** (1971), 4354.

G. C. Papanicolaou

1971. *Wave propagation in a one-dimensional random medium*, SIAM J. Appl. Math. **21** (1971), 13–18.

G. C. Papanicolaou and J. B. Keller

1972. *Stochastic differential equations with applications to random harmonic oscillators and wave propagation in random media*, SIAM J. Appl. Math. **21** (1971), 287–305.

D. N. Payton and W. M. Visscher

1967. Phys. Rev. **154** (1967), 802.

S. O. Rice

1944. *Mathematical analysis of random noise*, Bell System Tech. J. **23** (1944), 282–332. MR **6,** 89.

1945. *Mathematical analysis of random noise*, Bell System Tech. J. **24** (1945), 46–156. MR **6,** 233.

H. Scher and M. Lax

1969. Bull. Amer. Phys. Soc. **14** (1969), 311; and in preparation.

1972. J. Non-Cryst. Solids **8/9** (1972), 497 [Proc. Fourth Internat. Conf. on Amorphous and Liquid Semiconductors, 1971].

1973. *Stochastic transport in a disordered solid.* I, II, Phys. Rev. **B7** (1973), 4491–4519.

H. Scher and R. Zallen

1970. J. Chem. Phys. **53** (1970), 3759.

H. Schmidt

1957. *Disordered one-dimensional crystals*, Phys. Rev. (2) **105** (1957), 425–441. MR **18,** 961.

L. Schwartz and H. Ehrenreich

1971. Ann. Phys. **64** (1971), 100.

V. K. S. Shante and S. Kirkpatrick
1971. Adv. Physics **10** (1971), 325.
H. Shiba
1971. Progr. Theoret. Phys. **16** (1971), 77.
A. J. F. Siegert
1963. *Functional integrals in statistical mechanics*, Statistical Physics (Brandeis Summer Institute, 1962), vol. 3, Benjamin, New York, 1963, pp. 159–180. MR **27** #4607.
P. Soven
1966. Phys. Rev. **151** (1966), 539.
1967. Phys. Rev. **156** (1967), 809.
1969. Phys. Rev. **178** (1969), 1136.
M. F. Sykes and J. W. Essam
1964a. *Exact critical percolation probabilities for site and bond problems in two dimensions*, J. Mathematical Phys. **5** (1964), 1117–1127. MR **29** #1977.
1964b. Phys. Rev. **133** (1964), A310.
D. W. Taylor
1967. Phys. Rev. **156** (1967), 1017.
D. G. Thomas, J. J. Hopfield and W. M. Augustyniak
1965. Phys. Rev. **140** (1965), A202.
D. E. Thornton
1971. Phys. Rev. **4B** (1971), 3371.
D. J. Thouless
1970. J. Phys. (London) **C3** (1970), 1559.
1972. J. Non-Cryst. Solids **8/9** (1972).
V. A. Vyssotsky, S. B. Gordon, H. L. Frisch and J. M. Hammersley
1961. *Critical percolation probabilities (bond problem)*, Phys. Rev. (2) **123** (1961), 1566–1567. MR **24** #B1939.
K. M. Watson
1957. *Applications of scattering by quantum-mechanical systems*, Phys. Rev. (2) **105** (1957), 1388–1398. MR **18,** 853.
1953. *Multiple scattering and the many-body-problem-applications to photomeson production in complex nuclei*, Phys. Rev. (2) **89** (1953), 575–587. MR **14,** 829.
F. Yonezawa
1964. Progr. Theoret. Phys. **31** (1964), 357.
1968. Progr. Theoret. Phys. **40** (1968), 734.
F. Yonezawa and T. Matsubara
1966. Progr. Theoret. Phys. **35** (1966), 357.
R. Zallen and H. Scher
1971. Phys. Rev. **4B** (1971), 4471.
J. M. Ziman
1968. J. Phys. **C1** (1968), 1532.
1969. J. Phys. (London) **C2** (1969), 1230.
J. Zittartz and J. S. Langer
1966. Phys. Rev. **148** (1966), 741.

City College of New York, City University of New York
Bell Laboratories, Murray Hill

Analysis of Some Stochastic Ordinary Differential Equations

J. A. Morrison and J. McKenna

1. **Introduction.** In this paper we review a number of results concerning stochastic ordinary differential equations. The equations we study can be written symbolically as

$$du/dx = \boldsymbol{f}(\boldsymbol{u}(x), N(x), x). \tag{1.1}$$

In (1.1), $\boldsymbol{u}(x)$ is an L-dimensional column vector, $N(x)$ is a sample function of a stochastic process, and $\boldsymbol{f}(\cdot, \cdot, \cdot)$ is a deterministic L-vector valued function of its arguments.

By a stochastic process [**1**] we mean an ensemble of functions $\{N(x, \omega)\}$, each of which maps $0 \leqq x < \infty$ into some space S. The sample functions are indexed by the parameter ω which is a point in some probability space Ω. The space Ω is equipped with a probability measure $P(\omega)$. Stochastic averages of functions of the sample functions are defined by

$$\langle F(N(x)) \rangle = \int_{\Omega} F(N(x, \omega))\, dP(\omega). \tag{1.2}$$

The symbol $\langle\ \rangle$ will be used throughout this paper to denote stochastic averages.

Each sample function $N(x, \omega)$ defines a vector valued sample function $\boldsymbol{u}(x, \omega)$ by means of equation (1.1) and suitable initial or boundary conditions. The ensemble of solution sample functions $\{\boldsymbol{u}(x, \omega) | \omega \in \Omega\}$ defines a new random process, the solution process, with the same underlying

AMS (MOS) subject classifications (1970). Primary 34F05, 60-02, 60H10; Secondary 60G35, 60J10.

probability space Ω. We will typically suppress all mention of the underlying probability space Ω and talk of random functions $N(x)$ and solutions $\boldsymbol{u}(x)$. Clearly equation (1.1) stands for an ensemble of equations.

Stochastic differential equations arise in a variety of contexts in physics, engineering and economics. Typically one is interested in calculating the stochastic averages of various functions of the solutions. Although there are very few cases where one can express the solution $\boldsymbol{u}(x)$ explicitly as a functional of $N(x)$, there are nevertheless many other cases where various stochastic averages involving $\boldsymbol{u}(x)$ can be calculated either exactly or approximately. This paper is devoted to the study of a number of such calculations. We will consider both initial value problems and boundary value problems. We will be particularly interested in testing the validity of approximation schemes by applying them to cases where the exact solution is known. We make no claim to inclusiveness, and focus mainly on work central to our own research interests. Our main interest is in methods of solution, and we pay no attention to existence theorems and little attention to the physics which generates the equations.

We commence in §2 by considering a number of cases where the solution sample functions can be constructed explicitly. The simplest case is of course a first-order linear equation. Explicit expressions are obtained for the expectation of the solution in cases involving either a Gaussian process or the random telegraph process. A special case of the transmission line equations for a single line is then studied. Finally, several special cases of the one-dimensional wave equation are discussed.

In §3, we consider linear vector equations containing a white noise coefficient, and apply the Itô calculus [**2**] to obtain an equation for the expected value of the solution. We then use the result to calculate the first- and second-order moments of the solutions of the one-dimensional wave equation, with white noise fluctuations in the dielectric constant. We also consider the coupled line equations for two modes travelling in the same direction [**3**], in the case of white noise coupling between the modes. We calculate the expected values of the solutions, and also the covariances of the mode transfer functions. The results agree with those obtained by Rowe [**4**] for the wave equation, and by Rowe and Young [**5**], [**6**] for the coupled line equations. They used a different approach, involving a matrix technique, which we discuss for the calculation of the expected values of the solutions of the coupled line equations.

In §4, we consider (not necessarily linear) vector differential equations involving a finite state Markov chain, for which the solution together with the Markov chain forms a joint Markov process. The forward and backward Kolmogorov equations for the transition probability density functions are given. The results are applied in §5 to linear matrix differential equations with Markov chain coefficients, and equations are obtained for calculating the moments and correlation functions of the solutions. The equations are linear, and are subject to prescribed initial conditions.

In §6 we consider the particular case of the one-dimensional wave equation, with a dielectric constant that is a function of a finite state Markov chain. We also discuss some related work of J. E. Molyneux [**7**], who considered the same equation with a countable state space Markov process, rather than a finite state one. When the Markov chain has a stationary transition mechanism, the linear equations for calculating the moments of the solutions have constant coefficients. Hence they may be solved by means of Laplace transforms, and we give some results in the general case for the first- and second-order moments. In [**3**], the authors considered the coupled line equations for two modes travelling in the same direction, in the case in which the coupling is a function of a finite state Markov chain, but we do not give any details here.

In §7, we consider some particular examples for the wave equation. In the first example the fluctuations in the dielectric constant are proportional to the random telegraph process, and explicit expressions are given for the Laplace transforms of the first- and second-order moments of the solutions. The authors [**8**] also considered the correlation functions, but the details are omitted here. In the second example, the fluctuations in the dielectric constant vary as a linear combination of two stochastically independent random telegraph processes, with different rates. Explicit expressions for the Laplace transforms of the first- and second-order moments were obtained in [**9**], but are not given here. However, we do give approximate expressions for the first- and second-order moments, which are valid in the case of weak fluctuations in the dielectric constant. We also briefly mention two examples considered by Molyneux [**7**].

In §8, we turn to approximate methods, instead of exact ones. We discuss a perturbation method, which was developed independently by J. B. Keller [**10**], [**11**] and by R. C. Bourret [**12**], [**13**], [**14**], for calculating the expectation of the solutions of linear stochastic equations which contain a small parameter. We apply the method to the calculation of the first-order moments of the solutions of the wave equation, for small, wide sense stationary fluctuations in the dielectric constant. Explicit expressions are obtained for the Laplace transforms of the first-order moments, and it is shown that the results are exact both for white noise fluctuations, and for fluctuations proportional to the random telegraph process. This result was established indirectly by Bourret [**15**] in the latter case. The results are not exact when the fluctuations in the dielectric constant vary as a linear combination of two stochastically independent random telegraph processes, but they are consistent. This is shown by approximately inverting the Laplace transforms of the first-order moments, for general small wide sense stationary fluctuations. These general results are consistent with the perturbation results obtained by G. C. Papanicolaou and J. B. Keller [**16**], using a two-variable procedure.

In concluding §8, we show that the use of the first-order moments to calculate the second-order moments, by a widely used technique, leads to badly incorrect results. In §9, we apply the method of Keller and Bourret directly to the differential equations satisfied by the products of the solutions of the wave equation, and show that this leads to the exact results for the second-order moments, both for white noise fluctuations, and for fluctuations proportional to the random telegraph process. A procedure is also described for calculating the correlation functions of the solutions of the wave equation, which gives the exact results for random telegraph fluctuations. Approximate expressions are obtained for the second-order moments, for general small wide sense stationary fluctuations, and an approximate relationship, expressing the correlation functions in terms of the first- and second-order moments, is derived. These approximate results are consistent with the perturbation results obtained by Papanicolaou and Keller [**16**], using a two-variable procedure.

In §10 we give a direct proof of Bourret's result [**15**] that, for linear differential equations with a random telegraph coefficient, his method gives the exact results for the first-order moments of the solutions. We do this with the help of the exact results in §5, taking the finite state Markov chain to be the random telegraph process. We also show that the applications of the method, as discussed in §9, to the calculation of the second-order moments, and the correlation functions, lead to the exact results in this case.

In §11, we apply a limit theorem of R. Z. Has'minskiĭ [**17**] to investigate the behavior of the solutions of the wave equation with small, bounded, wide sense stationary fluctuations in the dielectric constant. The backward equation corresponding to an associated limit process, which is a diffusion process, is derived, leading to a formulation for the expectation of a function of the fundamental solution set, on an appropriately large interval of the independent variable. A transformation of variables [**18**] is made which leads to separation of the variables in the backward equation. We briefly discuss the application [**18**] to the calculation of the mean power transmitted through a randomly stratified dielectric slab by a plane electromagnetic wave normally incident on it, the nonstochastic dielectric constants on both sides of the slab being generally different from the mean dielectric constant in the slab. We also discuss the calculation [**18**] of the slowly varying part of the mean of a general function of the fundamental solution set of the wave equation.

In connection with the above, we remark that an intuitive derivation of the formal results of Has′minskiĭ's theorem was obtained, in more generality, by M. Lax [**19**]. Also, a more direct approach to the calculation of the mean power transmitted through a randomly stratified dielectric slab, using the limit theorem of Has′minskiĭ, was made by Papanicolaou [**20**], in the case in which the dielectric constants on both sides of the slab are equal to the mean dielectric constant in the slab. His procedure was to derive a (nonlinear) differential equation for the complex reflection coefficient, as a function of the thickness of the slab, and to apply the limit theorem to that equation, in order to calculate the mean reflected power.

Earlier Morrison, Papanicolaou and Keller [**21**] had asymptotically calculated the mean transmitted power when the small fluctuations in the dielectric constant in the slab are given by the Ornstein-Uhlenbeck process or by the random telegraph process. Only the dielectric constant on the incident side of the slab was permitted to differ from the mean dielectric constant in the slab. They obtained a perturbation solution to the relevant backward Kolmogorov equations, and in addition to the mean transmitted power also calculated the first- and second-order moments of the fundamental solution set of the one-dimensional wave equation. Approximations to the first-order moments in the case of the Ornstein-Uhlenbeck process had been obtained previously by U. Frisch [**22**], using another method. The problem of the reflection of an electromagnetic wave from a randomly stratified dielectric slab has been formulated by M. Fibich and E. Helfand [**23**] in the case of white noise fluctuations, but they did not attempt to solve the resulting Fokker-Planck-Kolmogorov equation.

In §12, we conclude by applying the two-variable perturbation results of Papanicolaou and Keller [**16**] to the coupled line equations for two modes travelling in the same direction. The covariances of the mode transfer functions are calculated in the case of weak, zero mean, wide sense stationary coupling. It is also shown that perturbation results are valid in the case of strong coupling, if the correlation length is short.

2. **Exactly solvable stochastic differential equations.** We begin by considering some stochastic differential equations which can be solved explicitly. We start with the first-order equation on $0 \leqq x < \infty$:

$$du/dx + i\beta_0(1 + \varepsilon N(x))u = 0, \tag{2.1}$$

with the nonstochastic initial condition

$$u(0) = 1. \tag{2.2}$$

In (2.1), β_0 and ε are positive constants, $i = (-1)^{1/2}$ and $N(x)$ is a real stochastic process. The solution of (2.1) satisfying (2.2) is

$$u(x) = \exp\left\{-i\beta_0\left[x + \varepsilon\int_0^x N(\xi)\,d\xi\right]\right\}. \tag{2.3}$$

Then,

$$\langle u(x)\rangle = \exp(-i\beta_0 x)\left\langle\exp\left\{-i\beta_0\varepsilon\int_0^x N(\xi)\,d\xi\right\}\right\rangle, \tag{2.4}$$

$$\langle u(x)u^*(y)\rangle = \exp(-i\beta_0(x-y))\left\langle\exp\left\{-i\beta_0\varepsilon\int_y^x N(\xi)\,d\xi\right\}\right\rangle, \tag{2.5}$$

where * denotes complex conjugate, and higher-order correlation functions can be expressed similarly.

Note that, from (2.5),

$$\langle|u(x)|^2\rangle = 1. \tag{2.6}$$

Thus the first two moments of $u(x)$ can be determined explicitly for any $N(x)$ for which $\langle\exp\{-i\beta_0\varepsilon\int_y^x N(\xi)\,d\xi\}\rangle$ can be determined explicitly.

The simplest such case is when $N(x)$ is independent of x and is thus simply a stochastic variable. This case has been studied in considerable detail in the literature and we refer the reader to [**22**] for a discussion and bibliography.

Perhaps the next simplest case is when $N(x)$ is a Gaussian random process [**1**]. This case has also been studied in detail in the literature, and we again refer the reader to [**22**] for an extended discussion and bibliography. However, we reproduce some of the results here for subsequent use. If $N(x)$ is a zero mean, wide sense stationary Gaussian process,

$$\langle N(x)\rangle = 0, \qquad \langle N(x)N(y)\rangle = \Gamma(x-y), \tag{2.7}$$

then

$$\langle u(x)\rangle = \exp(-i\beta_0 x)\exp\left\{-\frac{\beta_0^2\varepsilon^2}{2}\int_0^x\int_0^x \Gamma(\xi-\eta)\,d\xi\,d\eta\right\}, \tag{2.8}$$

$$\langle u(x)u^*(y)\rangle = \exp(-i\beta_0(x-y)) \cdot\exp\left\{-\frac{\beta_0^2\varepsilon^2}{2}\int_0^{|x-y|}\int_0^{|x-y|}\Gamma(\xi-\eta)\,d\xi\,d\eta\right\}. \tag{2.9}$$

It is clear from (2.9) that $u(x)$ is also a wide sense stationary process. In the important special case

$$\Gamma(\xi) = e^{-2b|\xi|}, \tag{2.10}$$

we have

(2.11) $$\langle u(x)\rangle = \exp(-i\beta_0 x)\exp\left[-\frac{\beta_0^2\varepsilon^2}{4b^2}(2bx - 1 + e^{-2bx})\right],$$

with a similar expression for $\langle u(x)u^*(y)\rangle$. We note from (2.11) that $\langle u(x)\rangle \to 0$ exponentially as $x \to \infty$, while from (2.6), $\langle |u(x)|^2\rangle \equiv 1$. This difference in the asymptotic behavior as $x \to \infty$ between the first- and second-order moments is a common characteristic of the solutions of many stochastic differential equations.

Another interesting case is when $N(x) = T(x)$, the random telegraph process [**24**]. The random telegraph process is an ensemble of square wave functions $\{T(x)\}$ such that each sample function can assume only the values ± 1. For a fixed x, a sample function chosen at random will equal 1 with probability $\frac{1}{2}$ or -1 with probability $\frac{1}{2}$. The probability $p(n, x)$ of a given sample function changing sign n times in an interval of length x is given by the Poisson process

(2.12) $$p(n, x) = ((bx)^n/n!)e^{-bx} \qquad (n = 0, 1, 2, \ldots),$$

where b is the average number of changes per unit length. It is not hard to show that

(2.13) $$\langle T(x)\rangle = 0, \qquad \langle T(x)T(y)\rangle = e^{-2b|x-y|}.$$

If we define $\theta(x)$ as the solution of the stochastic differential equation

(2.14) $$d\theta(x)/dx = T(x),$$

which satisfies the initial condition

(2.15) $$\theta(0) = 0,$$

then from (2.4), with $N(x) = T(x)$,

(2.16) $$\langle u(x)\rangle = \exp(-i\beta_0 x)\langle \exp(-i\beta_0\varepsilon\theta(x))\rangle,$$

so determining $\langle u(x)\rangle$ is equivalent to calculating the characteristic function of $\theta(x)$. This can be done in several ways. The first is to introduce the probability density functions $\sigma_j(\theta, x), j = 1, 2$, defined by

(2.17) $$\sigma_j(\theta, x)\, d\theta = \mathrm{Prob}\{\theta \leqq \theta(x) \leqq \theta + d\theta, T(x) = (-1)^{j-1}\}.$$

It can be shown that these probability density functions are the solutions of

(2.18) $$\begin{aligned}\partial\sigma_1/\partial x + \partial\sigma_1/\partial\theta + b(\sigma_1 - \sigma_2) &= 0,\\ \partial\sigma_2/\partial x - \partial\sigma_2/\partial\theta + b(\sigma_2 - \sigma_1) &= 0,\end{aligned}$$

which satisfy the boundary conditions

(2.19) $$\sigma_j(\theta, 0) = \tfrac{1}{2}\delta(\theta).$$

This is a special case of a much more general result discussed in §4. The boundary conditions are a direct consequence of (2.15) and definition (2.17). Also, it follows from (2.17) that

$$\langle \exp(-i\beta_0 \varepsilon \theta(x)) \rangle = \sum_{j=1}^{2} \int_{-\infty}^{\infty} \exp(-i\beta_0 \varepsilon \theta) \sigma_j(\theta, x)\, d\theta. \tag{2.20}$$

Equations (2.18) with boundary conditions (2.19) can be solved explicitly for $0 \leqq x < \infty$, and, for $j = 1, 2$,

$$\begin{aligned} &\sigma_j(\theta, x) - \tfrac{1}{2} e^{-bx} \delta[x + (-1)^j \theta] \\ &= \frac{b}{4} e^{-bx} \Big\{ I_0[b(x^2 - \theta^2)^{1/2}] \\ &\quad + \frac{[x + (-1)^{j+1}\theta]}{(x^2 - \theta^2)^{1/2}} I_1[b(x^2 - \theta^2)^{1/2}] \Big\} H(x - \theta) H(x + \theta). \end{aligned} \tag{2.21}$$

In (2.21), $\delta(x)$ is the delta function, $I_0(x)$ and $I_1(x)$ are modified Bessel functions of the first kind, and $H(x)$ is the Heaviside function:

$$\begin{aligned} H(x) &= 1, \qquad x > 0, \\ H(x) &= 0, \qquad x < 0. \end{aligned} \tag{2.22}$$

McFadden [**25**] had earlier derived the expression for $\sum_{j=1}^{2} \sigma_j(\theta, x)$. If the expressions (2.21) for $\sigma_j(\theta, x)$ are substituted into (2.20) we obtain

$$\begin{aligned} &\langle \exp(-i\beta_0 \varepsilon \theta(x)) \rangle \\ &= e^{-bx} \Big\{ \cosh[(b^2 - \beta_0^2 \varepsilon^2)^{1/2} x] + \frac{b}{(b^2 - \varepsilon^2 \beta_0^2)^{1/2}} \sinh[(b^2 - \beta_0^2 \varepsilon^2)^{1/2} x] \Big\}. \end{aligned} \tag{2.23}$$

Then (2.16) and (2.23) yield the expression for $\langle u(x) \rangle$.

We can define transition probability density functions, $\rho_{jk}(\theta, x; \varphi, \xi)$, $j, k = 1, 2$, corresponding to (2.17) by

$$\begin{aligned} &\rho_{jk}(\theta, x; \varphi, \xi)\, d\theta \\ &= \operatorname{Prob}\{\theta \leqq \theta(x) \leqq \theta + d\theta, T(x) = (-1)^{j-1} | \theta(\xi) = \varphi, T(\xi) = (-1)^{k-1}\}. \end{aligned} \tag{2.24}$$

The quantities $\rho_{1k}(\theta, x; \varphi, \xi)$ and $\rho_{2k}(\theta, x; \varphi, \xi)$ are the solutions of equations (2.18), for $k = 1, 2$, which satisfy the initial conditions

$$\rho_{jk}(\theta, \xi; \varphi, \xi) = \delta_{jk} \delta(\theta - \varphi). \tag{2.25}$$

These equations can also be solved explicitly, and with the aid of their solutions, $\langle u(x) u^*(y) \rangle$ can be calculated explicitly. We do not give the results here. This example of using probability density functions is a special case of a general technique which is described in greater detail in §4.

We can also calculate $\langle \exp(-i\beta_0\varepsilon\theta(x))\rangle$ by using a method due to D. A. Darling [**26**]. From (2.14) and (2.15) it follows trivially that

$$\theta(x) = \int_0^x T(\xi)\,d\xi. \tag{2.26}$$

If $T(x)$ is a sample function which changes sign at the n points x_j, $0 < x_1 < x_2 < \cdots < x_n < x$, then, from (2.26),

$$\begin{aligned}\theta(x) &= [\operatorname{sgn} T(0)] \\ &\quad\cdot [x_1 - (x_2 - x_1) + (x_3 - x_2) - \cdots + (-1)^n(x - x_n)].\end{aligned} \tag{2.27}$$

The probability density of n points chosen at random in $[0, x]$ when arranged in order is $n!/x^n$ over the simplex $0 \leqq x_1 \leqq x_2 \leqq \cdots \leqq x_n \leqq x$ and zero elsewhere. The probability of $T(\xi)$ changing sign n times in $[0, x]$ is just $p(n, x)$, given by (2.12), and $T(0) = \pm 1$, each with probability $\frac{1}{2}$. Therefore we can write

$$\begin{aligned}\langle \exp(-i\beta_0\varepsilon\theta(x))\rangle &= \frac{e^{-bx}}{2}\sum_{k=1}^{2}\sum_{n=0}^{\infty} b^n \\ &\quad\cdot \int_0^x\int_0^{x_n}\cdots\int_0^{x_2} \exp\{(-1)^k i\beta_0\varepsilon \\ &\quad\cdot [x_1 - (x_2 - x_1) + \cdots + (-1)^n(x - x_n)]\}\,dx_1\cdots dx_n.\end{aligned} \tag{2.28}$$

In (2.28), the integrals in the term $n = 0$ are to be interpreted as $\exp\{(-1)^k i\varepsilon\beta_0 x\}$. We note that $e^{bx}\langle \exp(-i\beta_0\varepsilon\theta(x))\rangle$ is an infinite sum of convolution integrals. Hence taking the Laplace transform with respect to x we get

$$\begin{aligned}&\int_0^\infty e^{-sx}e^{bx}\langle \exp(-i\beta_0\varepsilon\theta(x))\rangle\,dx \\ &\qquad = \frac{1}{2}\sum_{k=1}^{2}\sum_{n=0}^{\infty}\frac{b^{2n}}{(s^2+\varepsilon^2\beta_0^2)^n(s+(-1)^{k+1}i\varepsilon\beta_0)} \\ &\qquad\quad + \frac{1}{2}\sum_{k=1}^{2}\sum_{n=0}^{\infty}\frac{b^{2n+1}}{(s^2+\beta_0^2\varepsilon^2)^{n+1}} \\ &\qquad = \frac{b+s}{s^2+\varepsilon^2\beta_0^2-b^2}.\end{aligned} \tag{2.29}$$

Upon taking the inverse transform of the right-hand side of (2.29) we obtain equation (2.23). The correlation function defined in (2.5) can also be evaluated explicitly with the aid of Darling's method when $N(x) = T(x)$.

It should be noted that if ε is sufficiently small, then, for all x so that

$0 \leqq \beta_0 x \leqq O(1/\varepsilon^2)$, it follows from either (2.11) or from (2.16) and (2.23) that

$$\langle u(x)\rangle \sim \exp\{-i\beta_0 x - \varepsilon^2\beta_0^2 x/2b\}. \tag{2.30}$$

These are special examples of a general limit theorem discussed in §11. This general theorem shows that for bounded $N(x)$ satisfying (2.7), and a certain strong mixing condition, the asymptotics are determined solely by the correlation function $\Gamma(x - y)$.

We next consider several second-order stochastic equations or systems for which we can write down exact results. We consider first the telegraphist equations in the frequency domain for a single transmission line [**27**]

$$\begin{aligned} d\mathscr{U}/dx &= -\mathscr{Z}(x)\mathscr{I}(x), \\ d\mathscr{I}/dx &= -\mathscr{Y}(x)\mathscr{U}(x), \end{aligned} \tag{2.31}$$

where $\mathscr{U}(x)$ and $\mathscr{I}(x)$ are the time Fourier transforms of the voltage and current in the line, and $\mathscr{Z}(x)$ and $\mathscr{Y}(x)$ are the impedence and admittance, respectively. We assume that $\mathscr{Z}(x)$ and $\mathscr{Y}(x)$ are stochastic processes such that

$$K = (\mathscr{Z}(x)/\mathscr{Y}(x))^{1/2} \tag{2.32}$$

is a nonstochastic constant, independent of x. Then the system (2.31) has the two solutions

$$\mathscr{U}_\pm(x) = \exp\left\{\mp \int_0^x \gamma(\xi)\,d\xi\right\}, \qquad \mathscr{I}_\pm(x) = \pm\frac{1}{K}\mathscr{U}_\pm(x), \tag{2.33}$$

where

$$\gamma(x) = (\mathscr{Z}(x)\mathscr{Y}(x))^{1/2} = \mathscr{Z}(x)/K = K\mathscr{Y}(x). \tag{2.34}$$

From the first part of this section, we see that $\langle\mathscr{U}_\pm(x)\rangle$ and $\langle\mathscr{I}_\pm(x)\rangle$ can be calculated explicitly when $\gamma(x)$ is a random variable, independent of x; a Gaussian random variable; or the random telegraph process. Although this seems like a rather contrived example, it turns out to be of considerable interest since many of the statistics in the time domain can be calculated exactly. We refer the reader to reference [**27**] for details. More examples involving the telegraphist (or coupled line) equations are considered in §§3 and 12.

Another class of exactly solvable equations is of the form

$$d^2u/dx^2 + K(x)u = 0, \tag{2.35}$$

where the sample functions $K(x)$ are piecewise constant, nonstochastic initial conditions are prescribed at $x = 0$,

$$u(0) = u_0, \qquad u'(0) = v_0, \tag{2.36}$$

and the solution is desired on $0 \leqq x < \infty$. First we consider an example studied by Kerner [**28**] in an investigation of the theory of alloys. We define $K(x)$ by

$$K(x) = \alpha_n, \qquad n - 1 < x < n, \qquad n = 1, 2, \ldots, \tag{2.37}$$

where the α_n are identically distributed, stochastically independent random variables. For $n - 1 \leqq x \leqq n$, define

$$\boldsymbol{M}_n(x) = \begin{bmatrix} \cos \alpha_n^{1/2}(x - n + 1) & \dfrac{1}{\alpha_n^{1/2}} \sin \alpha_n^{1/2}(x - n + 1) \\ -\alpha_n^{1/2} \sin \alpha_n^{1/2}(x - n + 1) & \cos \alpha_n^{1/2}(x - n + 1) \end{bmatrix} \tag{2.38}$$

and let

$$\boldsymbol{U}(x) = \begin{bmatrix} u(x) \\ u'(x) \end{bmatrix}. \tag{2.39}$$

Then, for $n - 1 \leqq x \leqq n$, $n = 1, 2, \ldots$, and $\boldsymbol{U}$ continuous,

$$\boldsymbol{U}(x) = \boldsymbol{M}_n(x)\boldsymbol{M}_{n-1}(n - 1) \cdots \boldsymbol{M}_1(1)\boldsymbol{U}(0). \tag{2.40}$$

Since the α_n are identically distributed and stochastically independent, we can write

$$\langle \boldsymbol{U}(x) \rangle = \langle \boldsymbol{M}_n(x) \rangle \boldsymbol{M}^{n-1}\boldsymbol{U}(0), \tag{2.41}$$

where

$$\boldsymbol{M} = \langle \boldsymbol{M}_n(n) \rangle, \tag{2.42}$$

which is clearly independent of n. By considering the Kronecker product [**29**] $\boldsymbol{U}(x) \times \boldsymbol{U}(x)$, the second-order moments can be calculated:

$$\langle \boldsymbol{U}(x) \times \boldsymbol{U}(x) \rangle = \langle \boldsymbol{M}_n(x) \times \boldsymbol{M}_n(x) \rangle \boldsymbol{N}^{n-1}[\boldsymbol{U}(0) \times \boldsymbol{U}(0)], \tag{2.43}$$

where

$$\boldsymbol{N} = \langle \boldsymbol{M}_n(n) \times \boldsymbol{M}_n(n) \rangle. \tag{2.44}$$

In the same fashion, moments and correlation functions of all orders can be calculated.

The authors [**8**] have shown that a combination of the matrix methods used by Kerner plus Darling's method can be used to calculate exactly various moments and correlation functions of the solutions of (2.35) when

$$K(x) = \beta_0^2(1 + \varepsilon T(x)), \tag{2.45}$$

where β_0 and ε are positive constants and $T(x)$ is the random telegraph process. Since we will derive these results by a different method in §7, we do not discuss the calculation here, but refer the reader to reference [**8**] for details.

3. **Equations involving white noise.** Another class of equations for which exact results can be obtained involves white noise. Consider the vector equation

$$dy/dx = Gy + Hy\zeta(x), \tag{3.1}$$

where G and H are constant matrices, and $\zeta(x)$ is Gaussian white noise, with zero mean and spectral density 1. It is supposed that $y(0)$ is non-stochastic. Since we are interested in calculating only the moments of the solution $y(x)$, we may use the Itô calculus, rather than the Fokker-Planck equation corresponding to the system (3.1). According to W. M. Wonham [**30**], the Itô differential equation corresponding to (3.1) is

$$dy = (G + \tfrac{1}{2}H^2)y\,dx + Hy\,dw, \tag{3.2}$$

where w is a Wiener process. The result is stated for real equations, but it may be verified that it holds for complex G, H and y, by considering the equations for the real and imaginary parts. The integral equation corresponding to (3.2) is

$$y(x) = y(0) + (G + \tfrac{1}{2}H^2)\int_0^x y(\xi)\,d\xi + H\int_0^x y(\xi)\,dw(\xi). \tag{3.3}$$

The second integral in (3.3) is an Itô integral, and it has the property that its stochastic average is zero [**1**]. Thus

$$\langle y(x)\rangle = y(0) + (G + \tfrac{1}{2}H^2)\int_0^x \langle y(\xi)\rangle\,d\xi, \tag{3.4}$$

or, in differential form,

$$d\langle y(x)\rangle/dx = (G + \tfrac{1}{2}H^2)\langle y(x)\rangle. \tag{3.5}$$

We will also make use of the Itô calculus for the product of differentials [**2**], namely

$$(dw)^2 = dx, \qquad dw\,dx = 0, \qquad (dx)^2 = 0. \tag{3.6}$$

The first example we consider is the wave equation

$$\frac{du_m}{dx} = v_m, \qquad \frac{dv_m}{dx} = -\beta_0^2[1 + \varepsilon(D_0)^{1/2}\zeta(x)]u_m \qquad (m = 1, 2), \tag{3.7}$$

subject to the initial conditions

$$u_1(0) = 1 = v_2(0), \qquad u_2(0) = 0 = v_1(0), \tag{3.8}$$

where β_0, D_0 and ε are positive constants. The system (3.7) may be written in the form (3.1), with $\boldsymbol{y} = \operatorname{col}(u_m, v_m)$ and

$$\boldsymbol{G} = \begin{bmatrix} 0 & 1 \\ -\beta_0^2 & 0 \end{bmatrix}, \qquad \boldsymbol{H} = -\varepsilon\beta_0^2 D_0^{1/2} \begin{bmatrix} 0 & 0 \\ 1 & 0 \end{bmatrix}. \tag{3.9}$$

Then, setting

$$\boldsymbol{W}(x) = \begin{bmatrix} u_1(x) & u_2(x) \\ v_1(x) & v_2(x) \end{bmatrix}, \tag{3.10}$$

it follows from (3.2), since $\boldsymbol{H}^2 = \boldsymbol{0}$, that

$$d\boldsymbol{W} = (\boldsymbol{G}\,dx + \boldsymbol{H}\,dw)\boldsymbol{W}. \tag{3.11}$$

Also, from (3.5),

$$d\langle \boldsymbol{W}(x)\rangle/dx = \boldsymbol{G}\langle \boldsymbol{W}(x)\rangle. \tag{3.12}$$

Thus, the first-order moments are given by the solutions to (3.7), subject to (3.8), with $\varepsilon = 0$.

We consider the second-order moments next, and note that

$$d(\boldsymbol{W} \times \boldsymbol{W}) = (d\boldsymbol{W} \times \boldsymbol{W}) + (\boldsymbol{W} \times d\boldsymbol{W}) + (d\boldsymbol{W} \times d\boldsymbol{W}), \tag{3.13}$$

where $\times$ denotes Kronecker product $[\boldsymbol{A} \times \boldsymbol{B} = (a_{ij}) \times \boldsymbol{B} = (a_{ij}\boldsymbol{B})]$. From (3.6), (3.11) and (3.13), it follows that

$$\begin{aligned} d(\boldsymbol{W} \times \boldsymbol{W}) = {} & [(\boldsymbol{G} \times \boldsymbol{I}) + (\boldsymbol{I} \times \boldsymbol{G}) + (\boldsymbol{H} \times \boldsymbol{H})](\boldsymbol{W} \times \boldsymbol{W})\,dx \\ & + [(\boldsymbol{H} \times \boldsymbol{I}) + (\boldsymbol{I} \times \boldsymbol{H})](\boldsymbol{W} \times \boldsymbol{W})\,dw, \end{aligned} \tag{3.14}$$

where $\boldsymbol{I}$ is the unit matrix of order 2. Hence, corresponding to the relationship between (3.2) and (3.5),

$$d\langle \boldsymbol{W} \times \boldsymbol{W}\rangle/dx = [(\boldsymbol{G} \times \boldsymbol{I}) + (\boldsymbol{I} \times \boldsymbol{G}) + (\boldsymbol{H} \times \boldsymbol{H})]\langle \boldsymbol{W} \times \boldsymbol{W}\rangle. \tag{3.15}$$

From (3.8) and (3.10), the initial condition is $\boldsymbol{W}(0) = \boldsymbol{I}$. We denote the Laplace transform of a matrix $\boldsymbol{F}(x)$ by

$$\mathscr{L}(\boldsymbol{F}) = \int_0^\infty e^{-sx}\,\boldsymbol{F}(x)\,dx. \tag{3.16}$$

Then, from (3.9) and (3.15),

$$\begin{bmatrix} s & -1 & -1 & 0 \\ \beta_0^2 & s & 0 & -1 \\ \beta_0^2 & 0 & s & -1 \\ -\varepsilon^2\beta_0^4 D_0 & \beta_0^2 & \beta_0^2 & s \end{bmatrix} \mathscr{L}(\langle \boldsymbol{W} \times \boldsymbol{W}\rangle) = (\boldsymbol{I} \times \boldsymbol{I}). \tag{3.17}$$

Hence, inverting the matrix in (3.17),

$$[s(s^2 + 4\beta_0^2) - 2v]\mathscr{L}(\langle \boldsymbol{W} \times \boldsymbol{W} \rangle) \tag{3.18}$$

$$= \begin{bmatrix} (s^2 + 2\beta_0^2) & s & s & 2 \\ (v - \beta_0^2 s) & (s^2 + 2\beta_0^2 - v/s) & (v/s - 2\beta_0^2) & s \\ (v - \beta_0^2 s) & (v/s - 2\beta_0^2) & (s^2 + 2\beta_0^2 - v/s) & s \\ (2\beta_0^4 + vs) & (v - \beta_0^2 s) & (v - \beta_0^2 s) & (s^2 + 2\beta_0^2) \end{bmatrix}$$

where

$$v = \varepsilon^2 \beta_0^4 D_0. \tag{3.19}$$

The second example we consider is the coupled line equations for two modes travelling in the same direction [**3**],

$$\begin{aligned} dI_0/dx + \gamma_0 I_0(x) &= ic(x)I_1(x), \\ dI_1/dx + \gamma_1 I_1(x) &= ic(x)I_0(x), \end{aligned} \tag{3.20}$$

subject to the initial conditions

$$I_0(0) = i_0, \qquad I_1(0) = i_1. \tag{3.21}$$

Here $c(x)$ is a real coupling coefficient, which we take to be a random function of x, and the loss and phase constants are given by

$$\gamma_0 = \alpha_0 + i\beta_0, \qquad \gamma_1 = \alpha_1 + i\beta_1. \tag{3.22}$$

These equations provide an approximate description of a variety of physical systems [**31**], [**32**], [**33**] such as optical fibers [**34**], [**35**] and metal waveguides [**36**], [**37**]. Typically, with the choice $i_0 = 1$ and $i_1 = 0$, $I_0(x)$ represents a desired mode launched at $x = 0$, and $I_1(x)$ an undesired mode.

The differential loss and phase constants are given by

$$\Delta\gamma = \gamma_0 - \gamma_1 = \Delta\alpha + i\Delta\beta. \tag{3.23}$$

We write the coupling coefficient in the form

$$c(x) = C(\Delta\beta)N(x), \tag{3.24}$$

where C is assumed to be an odd function of $\Delta\beta$, and $N(x)$ is dimensionless. Let

$$g_0(x, \Delta\beta) = \exp(\gamma_0 x)I_0(x), \qquad g_1(x, \Delta\beta) = \exp(\gamma_0 x)I_1(x). \tag{3.25}$$

Then, from (3.20) and (3.23)–(3.25),

$$\begin{aligned} dg_0/dx &= iC(\Delta\beta)N(x)g_1(x, \Delta\beta), \\ dg_1/dx &= \Delta\gamma g_1(x, \Delta\beta) + iC(\Delta\beta)N(x)g_0(x, \Delta\beta), \end{aligned} \tag{3.26}$$

with initial conditions, from (3.21),

$$g_0(0, \Delta\beta) = i_0, \qquad g_1(0, \Delta\beta) = i_1. \tag{3.27}$$

We now define the correlation functions

$$R_{kl}(x) = \langle g_k(x, \Delta\beta + \sigma) g_l^*(x, \Delta\beta)\rangle \qquad (k, l = 0, 1), \tag{3.28}$$

where * denotes complex conjugate. We remark that the average powers in the two modes are given by

$$\begin{aligned} P_0(x) &= \exp(-2\alpha_0 x) R_{00}(x)|_{\sigma=0}, \\ P_1(x) &= \exp(-2\alpha_0 x) R_{11}(x)|_{\sigma=0}. \end{aligned} \tag{3.29}$$

H. E. Rowe and D. T. Young [**5**], [**6**] have calculated $\langle I_0(x)\rangle$, $\langle I_1(x)\rangle$ and $R_{kl}(x)$ for white noise coupling, in the case $i_0 = 1$ and $i_1 = 0$. They interpreted equations (3.20) and (3.21), which are symbolic when $c(x)$ is white noise, by means of a limiting process, and obtained their results with the aid of a matrix technique. We will discuss this technique at the end of the section, but first show that we may obtain their results by means of the Itô calculus, as we did in [**3**]. This shows that their interpretation of the equations in the case of white noise coupling is consistent with that of Stratonovič [**38**]. It is of interest to mention that Rowe and Young have also calculated the average of the squared envelope of the impulse response, which involves a double integral with respect to $\Delta\beta$ and σ of $R_{00}(x)$. Their calculation was exact for frequency-independent coupling [**5**], i.e., $C = c_0 \operatorname{sgn} \Delta\beta$, and approximate for narrow fractional bandwidth (small σ) for frequency-dependent coupling [**6**].

We now set

$$N(x) = D_0^{1/2}\zeta(x) \tag{3.30}$$

in (3.24), where $\zeta(x)$ is Gaussian white noise, with zero mean and spectral density 1. Then, from (3.20) and (3.24), corresponding to the relationship between (3.1) and (3.5),

$$d\langle I_l(x)\rangle/dx = -(\gamma_l + \tfrac{1}{2}C^2 D_0)\langle I_l(x)\rangle \qquad (l = 0, 1). \tag{3.31}$$

These equations agree with those obtained by Rowe and Young [**5**]. Also, from (3.1), (3.2), (3.26) and (3.30),

$$\begin{aligned} dg_0 &= -\tfrac{1}{2}C^2 D_0 g_0\, dx + iCD_0^{1/2} g_1\, dw, \\ dg_1 &= (\Delta\gamma - \tfrac{1}{2}C^2 D_0) g_1\, dx + iCD_0^{1/2} g_0\, dw. \end{aligned} \tag{3.32}$$

We define

$$r_{kl}(x) = g_k(x, \Delta\beta + \sigma) g_l^*(x, \Delta\beta) \qquad (k, l = 0, 1), \tag{3.33}$$

and let

$$C_0 = C(\Delta\beta), \qquad C_\sigma = C(\Delta\beta + \sigma). \tag{3.34}$$

Then, from (3.6) and (3.32),

$$\begin{aligned} dr_{00} &= D_0[C_0 C_\sigma r_{11} - \tfrac{1}{2}(C_0^2 + C_\sigma^2) r_{00}]\, dx \\ &\quad + iD_0^{1/2}(C_\sigma r_{10} - C_0 r_{01})\, dw, \end{aligned} \tag{3.35}$$

$$\begin{aligned} dr_{01} &= [\Delta\gamma^* - \tfrac{1}{2}(C_0^2 + C_\sigma^2) D_0] r_{01}\, dx + C_0 C_\sigma D_0 r_{10}\, dx \\ &\quad + iD_0^{1/2}(C_\sigma r_{11} - C_0 r_{00})\, dw, \end{aligned} \tag{3.36}$$

$$\begin{aligned} dr_{10} &= [(\Delta\gamma + i\sigma) - \tfrac{1}{2}(C_0^2 + C_\sigma^2) D_0] r_{10}\, dx + C_0 C_\sigma D_0 r_{01}\, dx \\ &\quad + iD_0^{1/2}(C_\sigma r_{00} - C_0 r_{11})\, dw, \end{aligned} \tag{3.37}$$

and

$$\begin{aligned} dr_{11} &= [(\Delta\gamma^* + \Delta\gamma + i\sigma) - \tfrac{1}{2}(C_0^2 + C_\sigma^2) D_0] r_{11}\, dx \\ &\quad + C_0 C_\sigma D_0 r_{00}\, dx + iD_0^{1/2}(C_\sigma r_{01} - C_0 r_{10})\, dw. \end{aligned} \tag{3.38}$$

From (3.28) and (3.33)–(3.38), corresponding to the relationship between (3.2) and (3.5), it follows that

$$dR_{00}/dx = -\tfrac{1}{2}(C_0^2 + C_\sigma^2) D_0 R_{00} + C_0 C_\sigma D_0 R_{11}, \tag{3.39}$$

$$dR_{11}/dx = C_0 C_\sigma D_0 R_{00} + [(\Delta\gamma^* + \Delta\gamma + i\sigma) - \tfrac{1}{2}(C_0^2 + C_\sigma^2) D_0] R_{11}, \tag{3.40}$$

and

$$dR_{01}/dx = [\Delta\gamma^* - \tfrac{1}{2}(C_0^2 + C_\sigma^2) D_0] R_{01} + C_0 C_\sigma D_0 R_{10}, \tag{3.41}$$

$$dR_{10}/dx = C_0 C_\sigma D_0 R_{01} + [(\Delta\gamma + i\sigma) - \tfrac{1}{2}(C_0^2 + C_\sigma^2) D_0] R_{10}. \tag{3.42}$$

These equations for R_{kl} are identical with those given by Rowe and Young [**5**], [**6**]. From (3.27) and (3.28), the initial conditions are

$$R_{kl}(0) = i_k i_l^*, \tag{3.43}$$

and the solutions are easy to write down.

We conclude this section by discussing the matrix method used by Rowe and Young [**5**], [**6**] for calculating the statistics of the solutions of the coupled line equations. Rowe has also applied this method to the one-dimensional wave equation [**4**], and earlier to the problem of random imperfections in waveguides [**37**]. We will illustrate here only the calculation of $\langle I_0(x)\rangle$ and $\langle I_1(x)\rangle$, when the coupling coefficient $c(x)$ in (3.20) is white noise. Rowe and Young approximated the line by a discrete model, obtained by dividing the line up into sections of length Δx, and approximating $c(x)$ by a sum of delta functions,

$$c_\delta(x) = \sum_j c_j \delta(x - j\Delta x + 0), \tag{3.44}$$

where

$$c_j = \int_{(j-1)\Delta x}^{j\Delta x} c(\xi)\, d\xi. \tag{3.45}$$

The approximation becomes exact in the limit $\Delta x \to 0$.

The different sections are independent since $c(x)$ is assumed to be white noise. The jth section of the discrete model consists of an ideal line section of length Δx, with zero coupling, terminated in a discrete (δ-function) mode converter of magnitude c_j. From (3.20), the ideal section satisfies

$$\begin{bmatrix} I_0(j\Delta x-) \\ I_1(j\Delta x-) \end{bmatrix} = \begin{bmatrix} \exp(-\gamma_0\Delta x) & 0 \\ 0 & \exp(-\gamma_1\Delta x) \end{bmatrix} \begin{bmatrix} I_0((j-1)\Delta x) \\ I_1((j-1)\Delta x) \end{bmatrix}. \tag{3.46}$$

Also, for the mode converter,

$$\begin{bmatrix} I_0(j\Delta x) \\ I_1(j\Delta x) \end{bmatrix} = \begin{bmatrix} \cos c_j & i \sin c_j \\ i \sin c_j & \cos c_j \end{bmatrix} \begin{bmatrix} I_0(j\Delta x-) \\ I_1(j\Delta x-) \end{bmatrix}. \tag{3.47}$$

Hence, combining (3.46) and (3.47),

$$\begin{aligned} \begin{bmatrix} I_0(j\Delta x) \\ I_1(j\Delta x) \end{bmatrix} &= \begin{bmatrix} \cos c_j \exp(-\gamma_0\Delta x) & i \sin c_j \exp(-\gamma_1\Delta x) \\ i \sin c_j \exp(-\gamma_0\Delta x) & \cos c_j \exp(-\gamma_1\Delta x) \end{bmatrix} \\ &\quad \cdot \begin{bmatrix} I_0((j-1)\Delta x) \\ I_1((j-1)\Delta x) \end{bmatrix}. \end{aligned} \tag{3.48}$$

With $c(x) = CD_0^{1/2}\zeta(x)$, where $\zeta(x)$ is Gaussian white noise, with zero mean and spectral density 1, it follows from (3.45) that the c_j are independent Gaussian random variables with

$$\langle c_j \rangle = 0, \qquad \langle c_j^2 \rangle = C^2 D_0 \Delta x, \qquad \langle c_j c_k \rangle = 0, \qquad j \neq k. \tag{3.49}$$

Thus, for small Δx,

$$\langle \sin c_j \rangle = 0, \qquad \langle \cos c_j \rangle = 1 - \tfrac{1}{2}C^2 D_0 \Delta x + O[(\Delta x)^2]. \tag{3.50}$$

Hence, taking expected values in equation (3.48), and making use of the independence of adjacent sections of the line, it follows that

$$\langle I_0(j\Delta x)\rangle = \{1 - (\gamma_0 + \tfrac{1}{2}C^2 D_0)\Delta x + O[(\Delta x)^2]\}\langle I_0((j-1)\Delta x)\rangle, \tag{3.51}$$

$$\langle I_1(j\Delta x)\rangle = \{1 - (\gamma_1 + \tfrac{1}{2}C^2 D_0)\Delta x + O[(\Delta x)^2]\}\langle I_1((j-1)\Delta x)\rangle. \tag{3.52}$$

Equation (3.31) is obtained in the limit $\Delta x \to 0$. Equations (3.39)-(3.42) for the quantities $R_{kl}(x)$, defined by (3.25) and (3.28), were obtained by Rowe and Young [**5**], [**6**] in an analogous manner.

4. **Vector differential equations involving Markov chains.** We now consider the real (not necessarily linear) vector differential equation

$$du/dx = \boldsymbol{f}(\boldsymbol{u}(x), M(x), x), \tag{4.1}$$

subject to the initial condition

$$\boldsymbol{u}(0) = \boldsymbol{g}[M(0)], \tag{4.2}$$

where $M(x)$ is a continuous parameter, finite state Markov chain [**1**] which has, in general, a nonstationary transition mechanism. Here $\boldsymbol{u}(x)$ is a column vector with components $u_i(x)$ $(i = 1, \ldots, L)$, and $\boldsymbol{f}(\cdot, \cdot, \cdot)$ and $\boldsymbol{g}(\cdot)$ are deterministic L-vector valued functions of their arguments. Applications in control theory involving equations of the form (4.1) have been considered by W. M. Wonham [**38**], and earlier by N. N. Krasovskiĭ and E. A. Lidskiĭ [**39**], [**40**], in the case that $M(x)$ has a stationary transition mechanism, and the initial vector $\boldsymbol{u}(0)$ does not depend on $M(0)$. However, our interest differs somewhat from theirs, and in the next section our results will be applied to obtain equations for the moments and correlation functions of the solutions of some linear matrix differential equations whose coefficients are functions of $M(x)$.

The sample functions $M(x)$ are defined on the half line $0 \leqq x < \infty$, can take on only a finite number N of distinct values a_p $(p = 1, \ldots, N)$, and have right continuous paths. An initial probability distribution is given

$$\alpha_p = \text{Prob}\{M(0) = a_p\} \qquad (p = 1, \ldots, N), \tag{4.3}$$

where $\alpha_p > 0$ and

$$\sum_{p=1}^{N} \alpha_p = 1. \tag{4.4}$$

We consider only those processes which can be defined by means of continuous, bounded infinitesimal generators. Thus we assume that there is given an $N \times N$ matrix function

$$\tau(x) = (\tau_{pq}(x)), \tag{4.5}$$

with the following properties holding for $0 \leqq x < \infty$:

$$\tau_{pq}(x) \geqq 0, \quad p \neq q; \qquad \tau_{pp}(x) \leqq 0 \qquad (p, q = 1, \ldots, N), \tag{4.6}$$

and

$$\sum_{q=1}^{N} \tau_{pq}(x) = 0 \qquad (p = 1, \ldots, N), \tag{4.7}$$

where $\tau_{pq}(x)$ is continuous and uniformly bounded.

Let $E_p^{(n)}(x, y)$ denote the event that $M(y) = a_p$ and $M(z)$ changes value n times in the interval (x, y). The events $E_p^{(n)}(x, y)$ and $E_p^{(m)}(x, y)$ are clearly mutually exclusive for $m \neq n$, and $\bigcup_{m=0}^{\infty} E_p^{(m)}(x, y)$ is just the event $M(y) = a_p$. We introduce the conditional probability

$$P_{qp}^{(n)}(x, y) = \mathrm{Prob}\{E_p^{(n)}(x, y) | M(x) = a_q\}. \tag{4.8}$$

Then there is the following probabilistic interpretation of the matrix $\tau(x)$. For $\delta x \to 0+$, and $p, q = 1, \ldots, N$, we have

$$P_{pp}^{(0)}(x, x + \delta x) = 1 + \tau_{pp}(x)\delta x + o(\delta x), \tag{4.9}$$

$$P_{qp}^{(1)}(x, x + \delta x) = \tau_{qp}(x)\delta x + o(\delta x), \qquad q \neq p, \tag{4.10}$$

and

$$\sum_{n=2}^{\infty} P_{qp}^{(n)}(x, x + \delta x) = o(\delta x). \tag{4.11}$$

If the matrix τ is constant then the process $M(x)$ is said to have a stationary transition mechanism.

We assume that sufficient conditions are imposed on $\boldsymbol{f}(\cdot, \cdot, \cdot)$ to ensure, for each sample function $M(x)$, the existence and uniqueness of the solution $\boldsymbol{u}(x)$ of (4.1) and (4.2) on some interval of the half line $0 \leqq x < \infty$ which contains the origin. It is easy to see that the joint process $\{\boldsymbol{u}(x), M(x)\}$ is a Markov process on this interval. Thus $\boldsymbol{u}(z)$ is determined uniquely, for $0 \leqq y \leqq z$, by $\boldsymbol{u}(y)$ and by $M(x)$ for $y \leqq x \leqq z$. But, since $M(x)$ is a Markov process, $M(x)$ for $x \geqq y$ is independent of $M(x')$ for $x' < y$, when conditioned on $M(y)$. It follows that $\{\boldsymbol{u}(x), M(x)\}$ is independent, for $x \geqq y$, of the random variables $(\boldsymbol{u}(x'), M(x'))$, $x' < y$, when conditioned on $(\boldsymbol{u}(y), M(y))$.

The probability density functions $\sigma_p(\boldsymbol{u}, x)$ $(p = 1, \ldots, N)$ are defined by

$$\sigma_p(\boldsymbol{u}, x) d^L\boldsymbol{u} = \mathrm{Prob}\{\boldsymbol{u} \leqq \boldsymbol{u}(x) \leqq \boldsymbol{u} + d\boldsymbol{u}, M(x) = a_p\}, \tag{4.12}$$

where $d^L\boldsymbol{u}$ is the volume element

$$d^L\boldsymbol{u} = \prod_{i=1}^{L} du_i. \tag{4.13}$$

Here $\boldsymbol{v} \leqq \boldsymbol{w}$ means that the inequality holds component by component. Stochastic averages of functions of the form $F(\boldsymbol{u}(x), M(x), x)$ are given by

$$\langle F(\boldsymbol{u}(x), M(x), x)\rangle = \sum_{p=1}^{N} \int_{R^L} F(\boldsymbol{u}, a_p, x)\sigma_p(\boldsymbol{u}, x)\, d^L\boldsymbol{u}, \tag{4.14}$$

the integration being over the entire Euclidean L-space, R^L. We remark that, for each $0 \leqq x < \infty$ and $p = 1, \ldots, N$, there is a positive probability that $M(\xi) \equiv a_p$ for $0 \leqq \xi \leqq x$. It follows that the $\sigma_p(\boldsymbol{u}, x)$ are generalized functions.

A formal derivation of the partial differential equations of which the probability density functions $\sigma_p(\boldsymbol{u}, x)$ are the (weak) solutions was given in [**41**], with the help of the relations (4.9)–(4.11) satisfied by the conditional probabilities (4.8). It was shown that

$$\frac{\partial \sigma_p}{\partial x} + \sum_{i=1}^{L} \frac{\partial}{\partial u_i}[f_i(\boldsymbol{u}, a_p, x)\sigma_p(\boldsymbol{u}, x)] - \sum_{r=1}^{N} \tau_{rp}(x)\sigma_r(\boldsymbol{u}, x) = 0. \tag{4.15}$$

From (4.2) and (4.3) the initial conditions are

$$\sigma_p(\boldsymbol{u}, 0) = \alpha_p \delta[\boldsymbol{u} - \boldsymbol{g}(a_p)] \qquad (p = 1, \dots, N), \tag{4.16}$$

where δ denotes the delta function. We remark that U. Frisch [**22**] has considered equation (4.1) in the case that $M(x)$ is a Markov diffusion process, rather than a finite chain, and he gave the corresponding Fokker-Planck equation for the probability density function. Equation (4.15) is the discrete version of this Fokker-Planck equation.

Since we are interested in the calculation of correlation functions, we also consider the transition probability density functions $\rho_{pq}(\boldsymbol{u}, x; \boldsymbol{v}, \xi)$ $(p, q = 1, \dots, N)$ defined for $0 \leqq \xi \leqq x$ by

$$\begin{aligned} &\rho_{pq}(\boldsymbol{u}, x; \boldsymbol{v}, \xi)\, d^L\boldsymbol{u} \\ &\qquad = \operatorname{Prob}\{\boldsymbol{u} \leqq \boldsymbol{u}(x) \leqq \boldsymbol{u} + d\boldsymbol{u}, M(x) = a_p | \boldsymbol{u}(\xi) = \boldsymbol{v}, M(\xi) = a_q\}. \end{aligned} \tag{4.17}$$

Stochastic averages of functions of the form $G(\boldsymbol{u}(x), M(x), x; \boldsymbol{u}(\xi), M(\xi), \xi)$ are given by

$$\begin{aligned} &\langle G(\boldsymbol{u}(x), M(x), x; \boldsymbol{u}(\xi), M(\xi), \xi)\rangle \\ &\qquad = \sum_{p=1}^{N} \sum_{q=1}^{N} \int_{R^L} \int_{R^L} G(\boldsymbol{u}, a_p, x; \boldsymbol{v}, a_q, \xi)\rho_{pq}(\boldsymbol{u}, x; \boldsymbol{v}, \xi)\sigma_q(\boldsymbol{v}, \xi)\, d^L\boldsymbol{u}\, d^L\boldsymbol{v}. \end{aligned} \tag{4.18}$$

Averages of functions involving three or more points can also be calculated with the aid of the transition probability density functions, since the joint process $\{\boldsymbol{u}(x), M(x)\}$ is a Markov process. For fixed q, $\boldsymbol{v}$ and ξ, the transition probability density functions $\rho_{pq}(\boldsymbol{u}, x; \boldsymbol{v}, \xi)$ satisfy the same partial differential equations as the probability density functions, namely

$$\begin{aligned} &\frac{\partial \rho_{pq}}{\partial x} + \sum_{i=1}^{L} \frac{\partial}{\partial u_i}[f_i(\boldsymbol{u}, a_p, x)\rho_{pq}(\boldsymbol{u}, x; \boldsymbol{v}, \xi)] \\ &\qquad - \sum_{r=1}^{N} \tau_{rp}(x)\rho_{rq}(\boldsymbol{u}, x; \boldsymbol{v}, \xi) = 0, \qquad x > \xi \geqq 0. \end{aligned} \tag{4.19}$$

However, from (4.17), the initial conditions are

$$\rho_{pq}(\boldsymbol{u}, \xi; \boldsymbol{v}, \xi) = \delta_{pq}\delta(\boldsymbol{u} - \boldsymbol{v}) \qquad (p, q = 1, \ldots, N), \tag{4.20}$$

where δ_{pq} is a Kronecker delta. It follows from (4.15) and (4.16), or from (4.2), (4.12) and (4.17), that

$$\sigma_p(\boldsymbol{u}, x) = \sum_{q=1}^{N} \alpha_q \rho_{pq}(\boldsymbol{u}, x; \boldsymbol{g}(a_q), 0). \tag{4.21}$$

Equations (4.19) are the forward equations for the transition probability density functions $\rho_{pq}(\boldsymbol{u}, x; \boldsymbol{v}, \xi)$. The corresponding backward equations, in which p, $\boldsymbol{u}$ and x are held fixed, are the adjoint equations

$$\frac{\partial \rho_{pq}}{\partial \xi} + \sum_{i=1}^{L} f_i(\boldsymbol{v}, a_q, \xi)\frac{\partial \rho_{pq}}{\partial v_i}(\boldsymbol{u}, x; \boldsymbol{v}, \xi) + \sum_{r=1}^{N} \tau_{qr}(\xi)\rho_{pr}(\boldsymbol{u}, x; \boldsymbol{v}, \xi) = 0, \qquad 0 \leqq \xi < x. \tag{4.22}$$

From (4.17) the boundary conditions are

$$\rho_{pq}(\boldsymbol{u}, x; \boldsymbol{v}, x) = \delta_{pq}\delta(\boldsymbol{u} - \boldsymbol{v}) \qquad (p, q = 1, \ldots, N). \tag{4.23}$$

We remark that in the case in which the process $M(x)$ has a stationary transition mechanism, Wonham [**38**] gives the infinitesimal generator for the joint process $\{\boldsymbol{u}(x), M(x)\}$, which was derived earlier by Krasovskiĭ and Lidskiĭ [**39**], [**40**]. The backward equation (4.22) may be derived by means of this generator. Now let

$$\mathscr{F}_q(x; \boldsymbol{v}, \xi) = \sum_{p=1}^{N} \int_{R^L} F(\boldsymbol{u}, a_p, x)\rho_{pq}(\boldsymbol{u}, x; \boldsymbol{v}, \xi)\, d^L\boldsymbol{u}. \tag{4.24}$$

Then, from (4.22),

$$\frac{\partial \mathscr{F}_q}{\partial \xi} + \sum_{i=1}^{L} f_i(\boldsymbol{v}, a_q, \xi)\frac{\partial \mathscr{F}_q}{\partial v_i} + \sum_{r=1}^{N} \tau_{qr}(\xi)\mathscr{F}_r = 0, \qquad 0 \leqq \xi < x, \tag{4.25}$$

with boundary conditions, from (4.23),

$$\mathscr{F}_q(x; \boldsymbol{v}, x) = F(\boldsymbol{v}, a_q, x) \qquad (q = 1, \ldots, N). \tag{4.26}$$

From (4.21) and (4.24) we obtain the following expression for the stochastic average in (4.14):

$$\langle F(\boldsymbol{u}(x), M(x), x)\rangle = \sum_{q=1}^{N} \alpha_q \mathscr{F}_q(x; \boldsymbol{g}(a_q), 0). \tag{4.27}$$

5. **Linear matrix differential equations with Markov chain coefficients.** We now apply the results of the previous section to the calculation of the moments and correlation functions of the solution of the stochastic matrix differential equation

(5.1) $$dW/dx = A(M(x), x)W(x),$$

satisfying the initial condition

(5.2) $$W(0) = \gamma[M(0)],$$

where $W(x)$ and $\gamma(\cdot)$ are $n \times m$ matrices and $A(\cdot, \cdot)$ is an $n \times n$ matrix valued function of its arguments. In component form we write

(5.3) $$W = (w_{ik}), \qquad \gamma = (\gamma_{ik}), \qquad A = (A_{ij}),$$

so that

(5.4) $$\frac{dw_{ik}}{dx} = \sum_{j=1}^{n} A_{ij}(M(x), x)w_{jk}(x), \qquad w_{ik}(0) = \gamma_{ik}[M(0)].$$

The system (5.4) may be written in vector form, by forming a column vector with nm components from the m successive column vectors $(w_{i1}), \ldots, (w_{im})$, with n components each. This leads to a system of the form (4.1) and (4.2), with $L = nm$. The probability density functions $\sigma_p(W, x)$ $(p = 1, \ldots, N)$ corresponding to the system (5.4) satisfy, from (4.15), the equations

(5.5) $$\frac{\partial \sigma_p}{\partial x} + \sum_{i=1}^{n} \sum_{j=1}^{n} \sum_{k=1}^{m} A_{ij}(a_p, x)\frac{\partial(w_{jk}\sigma_p)}{\partial w_{ik}} - \sum_{r=1}^{N} \tau_{rp}(x)\sigma_r = 0,$$

with initial conditions, from (4.16),

(5.6) $$\sigma_p(W, 0) = \alpha_p \delta[W - \gamma(a_p)].$$

Since equation (5.5) is homogeneous in the elements of W, it is possible to obtain equations for the moments of any given order. We denote by $W^{(s)}$ the s-fold Kronecker product of the matrix W with itself. Then the sth order moments of σ_p are given by

(5.7) $$\langle W(x)^{(s)} \rangle_p = \int_{R^{nm}} W^{(s)}\sigma_p(W, x)\, d^{nm}W,$$

where $d^{nm}W = \prod_{i=1}^{n} \prod_{k=1}^{m} dw_{ik}$. It follows from (4.14) that the expected value of $W^{(s)}$ is given by

(5.8) $$\langle W(x)^{(s)} \rangle = \sum_{p=1}^{N} \langle W(x)^{(s)} \rangle_p.$$

For the first-order moments it is found from (5.5), after some integrations by parts, that

(5.9) $$\frac{d}{dx}\langle W(x) \rangle_p = A(a_p, x)\langle W(x) \rangle_p + \sum_{r=1}^{N} \tau_{rp}(x)\langle W(x) \rangle_r,$$

with initial conditions, from (5.6),

$$\langle W(0)\rangle_p = \alpha_p \gamma(a_p) \qquad (p = 1, \ldots, N). \tag{5.10}$$

Similarly, for the second-order moments, it is found that, with I_n the unit matrix of order n,

$$d\langle W(x) \times W(x)\rangle_p/dx$$

$$= \sum_{r=1}^{N} \tau_{rp}(x)\langle W(x) \times W(x)\rangle_r \tag{5.11}$$
$$+ \{[A(a_p, x) \times I_n] + [I_n \times A(a_p, x)]\}\langle W(x) \times W(x)\rangle_p,$$

with initial conditions

$$\langle W(0) \times W(0)\rangle_p = \alpha_p[\gamma(a_p) \times \gamma(a_p)] \qquad (p = 1, \ldots, N). \tag{5.12}$$

Actually, equations (5.11) and (5.12) may be obtained by first deriving the equation satisfied by $W(x) \times W(x)$, and then applying the general result for first-order moments. Thus, from (5.1),

$$\begin{aligned} &\frac{d}{dx}[W(x) \times W(x)] \\ &\quad = \{[A(M(x), x) \times I_n] + [I_n \times A(M(x), x)]\}[W(x) \times W(x)], \end{aligned} \tag{5.13}$$

with initial condition, from (5.2),

$$W(0) \times W(0) = \gamma[M(0)] \times \gamma[M(0)]. \tag{5.14}$$

In this manner it is found that, generally,

$$\begin{aligned} \frac{d}{dx}\langle W(x)^{(s)}\rangle_p &= \sum_{r=1}^{N} \tau_{rp}(x)\langle W(x)^{(s)}\rangle_r \\ &\quad + \sum_{l=1}^{s} [I_n^{(l-1)} \times A(a_p, x) \times I_n^{(s-l)}]\langle W(x)^{(s)}\rangle_p, \end{aligned} \tag{5.15}$$

where $I_n^{(0)} = I_1$, with initial conditions

$$\langle W(0)^{(s)}\rangle_p = \alpha_p[\gamma(a_p)^{(s)}] \qquad (p = 1, \ldots, N). \tag{5.16}$$

Thus the expectation of $W^{(s)}$ may be calculated by solving the system of linear matrix differential equations (5.15), subject to the initial conditions (5.16), and then using (5.8). We remark that in the case in which the process $M(x)$ has a stationary transition mechanism, and A is a function of $M(x)$ alone, these equations have constant coefficients.

We next consider the calculation of the correlation functions. Corresponding to (5.5), the forward equations satisfied by the transition probability density functions $\rho_{pq}(W, x; V, \xi)\,(p, q = 1, \ldots, N)$ are, from (4.19), for $x > \xi \geqq 0$,

$$(5.17)\qquad \frac{\partial \rho_{pq}}{\partial x} + \sum_{i=1}^{n} \sum_{j=1}^{n} \sum_{k=1}^{m} A_{ij}(a_p, x) \frac{\partial (w_{jk}\rho_{pq})}{\partial w_{ik}} - \sum_{r=1}^{N} \tau_{rp}(x)\rho_{rq} = 0,$$

with initial conditions, from (4.20),

$$(5.18)\qquad \rho_{pq}(\boldsymbol{W}, \xi; \boldsymbol{V}, \xi) = \delta_{pq}\delta(\boldsymbol{W} - \boldsymbol{V}).$$

Let the matrices $\boldsymbol{\Psi}_{pq}(x, \xi)$ satisfy the equations

$$(5.19)\qquad \partial \boldsymbol{\Psi}_{pq}/\partial x = \boldsymbol{A}(a_p, x)\boldsymbol{\Psi}_{pq} + \sum_{r=1}^{N} \tau_{rp}(x)\boldsymbol{\Psi}_{rq},$$

with initial conditions

$$(5.20)\qquad \boldsymbol{\Psi}_{pq}(\xi, \xi) = \delta_{pq}\boldsymbol{I}_n \qquad (p, q = 1, \ldots, N).$$

Note, from (5.9) and (5.10), that

$$(5.21)\qquad \langle \boldsymbol{W}(x)\rangle_p = \sum_{q=1}^{N} \alpha_q \boldsymbol{\Psi}_{pq}(x, 0)\gamma(a_q).$$

It may be verified, from (5.17)–(5.20), that

$$(5.22)\qquad \int_{R^{nm}} \boldsymbol{W}\rho_{pq}(\boldsymbol{W}, x; \boldsymbol{V}, \xi)\, d^{nm}\boldsymbol{W} = \boldsymbol{\Psi}_{pq}(x, \xi)\boldsymbol{V}.$$

But, from (4.18), for $0 \leqq \xi \leqq x$,

$$(5.23)\qquad \begin{aligned} &\langle \boldsymbol{W}(x) \times \boldsymbol{W}(\xi)\rangle \\ &= \sum_{p=1}^{N} \sum_{q=1}^{N} \int_{R^{nm}} \int_{R^{nm}} (\boldsymbol{W} \times \boldsymbol{V})\rho_{pq}(\boldsymbol{W}, x; \boldsymbol{V}, \xi)\sigma_q(\boldsymbol{V}, \xi)\, d^{nm}\boldsymbol{W} d^{nm}\boldsymbol{V}. \end{aligned}$$

Hence, from (5.7) and (5.22), we obtain the following expression for the correlation functions, for $0 \leqq \xi \leqq x$,

$$(5.24)\qquad \langle \boldsymbol{W}(x) \times \boldsymbol{W}(\xi)\rangle = \sum_{q=1}^{N} \left[\sum_{p=1}^{N} \boldsymbol{\Psi}_{pq}(x, \xi) \times \boldsymbol{I}_n \right] \langle \boldsymbol{W}(\xi) \times \boldsymbol{W}(\xi)\rangle_q.$$

The equations satisfied by the various moments which we have just given are formally elegant, and are useful in obtaining relations such as equation (5.24); and as we will show in §10, they are useful for studying certain approximation schemes. However, an alternative formulation of the equations, which we will now give, often turns out to be more useful in the actual computation of the moments. This is because in this new formulation, the systems of equations for the moments often decouple into smaller subsystems, making the calculations easier. This will be illustrated by example in §6.

We begin by considering the first-order moments, and let the vector

$\omega_h(x)$ denote the hth row of the matrix $W(x)$, that is

(5.25) $$\omega_h(x) = (w_{h1}(x), \ldots, w_{hm}(x)) \qquad (h = 1, \ldots, n),$$

and let the matrix $\boldsymbol{\Omega}_h(x)$ denote the column of vectors

(5.26) $$\boldsymbol{\Omega}_h(x) = \operatorname{col}(\langle \omega_h(x) \rangle_1, \ldots, \langle \omega_h(x) \rangle_N).$$

Then, from (5.8), the expected value of ω_h is

(5.27) $$\langle \omega_h(x) \rangle = \boldsymbol{E}_N \boldsymbol{\Omega}_h(x),$$

where $\boldsymbol{E}_N$ is the row vector with all elements equal to 1. We define the $N \times N$ diagonal matrices $\boldsymbol{D}_{ij}(x)$ and $\boldsymbol{v}_{jk}$ $(i, j = 1, \ldots, n)$ $(k = 1, \ldots, m)$ by

(5.28) $$\boldsymbol{D}_{ij}(x) = \operatorname{diag}[A_{ij}(a_p, x)],$$

where p ranges from 1 to N along the diagonal, and

(5.29) $$\boldsymbol{v}_{jk} = \operatorname{diag}[\gamma_{jk}(a_p)].$$

Then (5.9) may be rewritten in the form

(5.30) $$d\boldsymbol{\Omega}_h/dx = \sum_{j=1}^{n} \boldsymbol{D}_{hj}(x)\boldsymbol{\Omega}_j(x) + \boldsymbol{\tau}^t(x)\boldsymbol{\Omega}_h(x),$$

where t denotes transpose. Also, from (5.10), the initial conditions are

(5.31) $$\boldsymbol{\Omega}_h(0) = (\boldsymbol{v}_{h1}\boldsymbol{\alpha}^t, \ldots, \boldsymbol{v}_{hm}\boldsymbol{\alpha}^t) \qquad (h = 1, \ldots, n),$$

where

(5.32) $$\boldsymbol{\alpha} = (\alpha_1, \ldots, \alpha_N)$$

is the row vector of initial probabilities given by (4.3). Note, from (4.4), that $\boldsymbol{E}_N\boldsymbol{\alpha}^t = 1$.

We consider next the second-order moments, and define the column of matrices

(5.33) $$\boldsymbol{\Omega}_{fh}(x) = \operatorname{col}(\langle \omega_f(x) \times \omega_h(x) \rangle_1, \ldots, \langle \omega_f(x) \times \omega_h(x) \rangle_N).$$

Then, from (5.25) and (5.28), equation (5.11) may be rewritten in the form

(5.34) $$\frac{d\boldsymbol{\Omega}_{fh}}{dx} = \boldsymbol{\tau}^t(x)\boldsymbol{\Omega}_{fh}(x) + \sum_{j=1}^{n} [\boldsymbol{D}_{fj}(x)\boldsymbol{\Omega}_{jh}(x) + \boldsymbol{D}_{hj}(x)\boldsymbol{\Omega}_{fj}(x)].$$

We define the $N \times N$ diagonal matrices ξ_{fghl} $(f, h = 1, \ldots, n)$ $(g, l = 1, \ldots, m)$ by

(5.35) $$\xi_{fghl} = \operatorname{diag}[\gamma_{fg}(a_p)\gamma_{hl}(a_p)].$$

Then, from (5.12), the initial conditions are

(5.36) $$\boldsymbol{\Omega}_{fh}(0) = (\xi_{f1h1}\boldsymbol{\alpha}^t, \ldots, \xi_{f1hm}\boldsymbol{\alpha}^t, \ldots, \xi_{fmh1}\boldsymbol{\alpha}^t, \ldots, \xi_{fmhm}\boldsymbol{\alpha}^t).$$

Finally, from (5.8), the expected value of $(\boldsymbol{\omega}_f \times \boldsymbol{\omega}_h)$ is

$$\langle \boldsymbol{\omega}_f(x) \times \boldsymbol{\omega}_h(x) \rangle = \boldsymbol{E}_N \boldsymbol{\Omega}_{fh}(x). \tag{5.37}$$

Analogous equations may be written down for the higher-order moments [**41**].

Now we consider the alternate formulation for the calculation of the correlation functions. We define the $N \times N$ matrix $\boldsymbol{\rho}$ and the column vector $\boldsymbol{\sigma}$ by

$$\boldsymbol{\rho} = (\rho_{pq}), \qquad \boldsymbol{\sigma} = \operatorname{col}(\sigma_1, \ldots, \sigma_N). \tag{5.38}$$

Then, from (5.28), equation (5.17) may be written in the form

$$\frac{\partial \boldsymbol{\rho}}{\partial x} + \sum_{i=1}^{n} \sum_{j=1}^{n} \sum_{k=1}^{m} \boldsymbol{D}_{ij}(x) \frac{\partial (w_{jk}\boldsymbol{\rho})}{\partial w_{ik}} - \boldsymbol{\tau}^t(x)\boldsymbol{\rho} = 0, \tag{5.39}$$

with initial condition, from (5.18),

$$\boldsymbol{\rho}(\boldsymbol{W}, \xi; \boldsymbol{V}, \xi) = \delta(\boldsymbol{W} - \boldsymbol{V})\boldsymbol{I}_N. \tag{5.40}$$

Also, from (4.18), for $0 \leqq \xi \leqq x$,

$$\begin{aligned} &\langle w_{fg}(\xi) w_{hl}(x) \rangle \\ &\quad = \boldsymbol{E}_N \int_{R^{nm}} \int_{R^{nm}} v_{fg} w_{hl} \boldsymbol{\rho}(\boldsymbol{W}, x; \boldsymbol{V}, \xi) \boldsymbol{\sigma}(\boldsymbol{V}, \xi)\, d^{nm}\boldsymbol{W}\, d^{nm}\boldsymbol{V}. \end{aligned} \tag{5.41}$$

Let the matrices $\boldsymbol{\theta}_{hk}(x, \xi)$ satisfy the equations

$$\partial \boldsymbol{\theta}_{hk}/\partial x = \sum_{j=1}^{n} \boldsymbol{D}_{hj}(x)\boldsymbol{\theta}_{jk} + \boldsymbol{\tau}^t(x)\boldsymbol{\theta}_{hk}, \tag{5.42}$$

with initial conditions

$$\boldsymbol{\theta}_{hk}(\xi, \xi) = \delta_{hk}\boldsymbol{I}_N \qquad (h, k = 1, \ldots, n). \tag{5.43}$$

Note, from (5.30), that

$$\boldsymbol{\Omega}_h(x) = \sum_{k=1}^{n} \boldsymbol{\theta}_{hk}(x, 0)\boldsymbol{\Omega}_k(0). \tag{5.44}$$

It may be verified, from (5.39), (5.40), (5.42) and (5.43), that

$$\int_{R^{nm}} w_{hl} \boldsymbol{\rho}(\boldsymbol{W}, x; \boldsymbol{V}, \xi)\, d^{nm}\boldsymbol{W} = \sum_{k=1}^{n} \boldsymbol{\theta}_{hk}(x, \xi) v_{kl}. \tag{5.45}$$

It follows, from (5.7), (5.25), (5.33), (5.38) and (5.41), that

$$\langle \boldsymbol{\omega}_f(\xi) \times \boldsymbol{\omega}_h(x) \rangle = \sum_{k=1}^{n} \boldsymbol{E}_N \boldsymbol{\theta}_{hk}(x, \xi)\boldsymbol{\Omega}_{fk}(\xi), \qquad 0 \leqq \xi \leqq x. \tag{5.46}$$

6. **Application to the wave equation.** We now apply the results of the previous section to the solutions of the equations

$$\frac{du_m}{dx} = v_m, \quad \frac{dv_m}{dx} = -\beta_0^2[1 + \varepsilon f(M(x))]u_m \quad (m = 1, 2), \tag{6.1}$$

which satisfy the nonstochastic initial conditions

$$u_1(0) = 1 = v_2(0), \quad u_2(0) = 0 = v_1(0). \tag{6.2}$$

The system (6.1) may be written in the matrix form (5.1) with

$$\boldsymbol{W} = \begin{bmatrix} u_1 & u_2 \\ v_1 & v_2 \end{bmatrix}, \quad \boldsymbol{A} = \begin{bmatrix} 0 & 1 \\ -\beta_0^2[1 + \varepsilon f(M(x))] & 0 \end{bmatrix}. \tag{6.3}$$

From (6.2) the initial condition is

$$\boldsymbol{W}(0) = \boldsymbol{I}_2. \tag{6.4}$$

Thus $n = 2, m = 2$ and, corresponding to (5.28) and (5.29),

$$\boldsymbol{D}_{11} = \boldsymbol{0} = \boldsymbol{D}_{22}, \quad \boldsymbol{D}_{12} = \boldsymbol{I}_N, \quad \boldsymbol{D}_{21} = -\boldsymbol{B}, \tag{6.5}$$

where

$$\boldsymbol{B} = \beta_0^2 \boldsymbol{I}_N + \varepsilon\beta_0^2 \operatorname{diag}[f(a_p)], \tag{6.6}$$

and

$$\boldsymbol{v}_{jk} = \delta_{jk}\boldsymbol{I}_N. \tag{6.7}$$

The system (6.1) and (6.2) was considered by the authors [**9**] who obtained equations for the first- and second-order moments of the solutions, which may be recovered from the general results. Thus, we define the four column vectors

$$\boldsymbol{U}_m(x) = (\langle u_m(x)\rangle_p), \quad \boldsymbol{V}_m(x) = (\langle v_m(x)\rangle_p) \quad (m = 1, 2), \tag{6.8}$$

where $p = 1, \ldots, N$. Then, from (5.25) and (5.26),

$$\boldsymbol{\Omega}_1 = (\boldsymbol{U}_1, \boldsymbol{U}_2), \quad \boldsymbol{\Omega}_2 = (\boldsymbol{V}_1, \boldsymbol{V}_2), \tag{6.9}$$

and from (5.27),

$$\langle u_m(x)\rangle = \boldsymbol{E}_N\boldsymbol{U}_m(x), \quad \langle v_m(x)\rangle = \boldsymbol{E}_N\boldsymbol{V}_m(x) \quad (m = 1, 2). \tag{6.10}$$

But, from (5.30) and (6.5),

$$\frac{d}{dx}\begin{bmatrix} \boldsymbol{U}_m(x) \\ \boldsymbol{V}_m(x) \end{bmatrix} = \begin{bmatrix} \boldsymbol{\tau}^t(x) & \boldsymbol{I}_N \\ -\boldsymbol{B} & \boldsymbol{\tau}^t(x) \end{bmatrix}\begin{bmatrix} \boldsymbol{U}_m(x) \\ \boldsymbol{V}_m(x) \end{bmatrix}, \tag{6.11}$$

with initial conditions, from (5.31) and (6.7),

$$U_1(0) = \alpha^t = V_2(0), \qquad U_2(0) = \mathbf{0} = V_1(0). \tag{6.12}$$

Notice that the equations for $m = 1$ and 2 are uncoupled.

Next we define the nine column vectors

$$X_m(x) = (X_m(x)_p), \qquad Y_m(x) = (Y_m(x)_p), \qquad Z_m(x) = (Z_m(x)_p), \tag{6.13}$$

for $m = 0, 1, 2$ and $p = 1, \ldots, N$, by

$$X_0(x)_p = \langle u_1(x)u_2(x)\rangle_p, \qquad X_m(x)_p = \langle u_m^2(x)\rangle_p \qquad (m = 1, 2), \tag{6.14}$$

$$Y_0(x)_p = \tfrac{1}{2}[\langle u_1(x)v_2(x)\rangle_p + \langle u_2(x)v_1(x)\rangle_p], \tag{6.15}$$

$$Y_m(x)_p = \langle u_m(x)v_m(x)\rangle_p \qquad (m = 1, 2), \tag{6.16}$$

$$Z_0(x)_p = \langle v_1(x)v_2(x)\rangle_p, \qquad Z_m(x)_p = \langle v_m^2(x)\rangle_p \qquad (m = 1, 2). \tag{6.17}$$

Then, from (5.25) and (5.33),

$$\mathbf{\Omega}_{11} = (X_1, X_0, X_0, X_2), \qquad \mathbf{\Omega}_{22} = (Z_1, Z_0, Z_0, Z_2), \tag{6.18}$$

and

$$\tfrac{1}{2}(\mathbf{\Omega}_{12} + \mathbf{\Omega}_{21}) = (Y_1, Y_0, Y_0, Y_2). \tag{6.19}$$

It follows from (5.34) and (6.5) that

$$\frac{d}{dx}\begin{bmatrix} X_m(x) \\ Y_m(x) \\ Z_m(x) \end{bmatrix} = \begin{bmatrix} \tau^t(x) & 2I_N & \mathbf{0} \\ -B & \tau^t(x) & I_N \\ \mathbf{0} & -2B & \tau^t(x) \end{bmatrix} \begin{bmatrix} X_m(x) \\ Y_m(x) \\ Z_m(x) \end{bmatrix}, \tag{6.20}$$

and

$$d[\mathbf{\Omega}_{12}(x) - \mathbf{\Omega}_{21}(x)]/dx = \tau^t(x)[\mathbf{\Omega}_{12}(x) - \mathbf{\Omega}_{21}(x)]. \tag{6.21}$$

Since, from (5.35),

$$\xi_{fghl} = \delta_{fg}\delta_{hl}I_N, \tag{6.22}$$

the initial conditions are, from (5.36),

$$X_0(0) = X_2(0) = Y_1(0) = Y_2(0) = Z_0(0) = Z_1(0) = \mathbf{0}, \tag{6.23}$$

$$X_1(0) = 2Y_0(0) = Z_2(0) = \alpha^t, \tag{6.24}$$

and

$$[\mathbf{\Omega}_{12}(0) - \mathbf{\Omega}_{21}(0)] = \alpha^t(0, 1, -1, 0). \tag{6.25}$$

Since, from (4.4) and (4.7),

$$E_N\alpha^t = 1, \qquad E_N\tau^t(x) \equiv 0, \tag{6.26}$$

it follows from (6.21) and (6.25) that

$$E_N[\mathbf{\Omega}_{12}(x) - \mathbf{\Omega}_{21}(x)] \equiv (0, 1, -1, 0), \tag{6.27}$$

which is equivalent to the identity

$$u_1(x)v_2(x) - u_2(x)v_1(x) \equiv 1. \tag{6.28}$$

From (5.25) and (5.35) the second-order moments are given by

$$\begin{gathered}\langle u_m^2(x)\rangle = E_N X_m(x), \qquad \langle v_m^2(x)\rangle = E_N Z_m(x),\\ \langle u_m(x)v_m(x)\rangle = E_N Y_m(x) \qquad (m = 1, 2),\end{gathered} \tag{6.29}$$

$$\langle u_1(x)u_2(x)\rangle = E_N X_0, \qquad \langle v_1(x)v_2(x)\rangle = E_N Z_0, \tag{6.30}$$

and, using (6.28),

$$\langle u_1(x)v_2(x)\rangle = E_N Y_0(x) + \tfrac{1}{2}, \qquad \langle u_2(x)v_1(x)\rangle = E_N Y_0(x) - \tfrac{1}{2}. \tag{6.31}$$

We remark that J. E. Molyneux [7] considered the system (6.1) and (6.2) with $M(x)$ replaced by a countable state space Markov process $\mathscr{M}(x)$, assuming values in a set $S = \{a_j : j \in J\}$, where J is either $Z^+ = \{1, 2, \ldots\}$ or $Z = \{\ldots, -2, -1, 0, 1, 2, \ldots\}$. (For notational convenience we do not include 0 in the first set.) In either case there are infinitely many equations to be solved in order to calculate the moments, of a given order, of the solutions. However, Molyneux considered two cases in which these equations may be reduced to a finite number. Thus, he assumed that

$$f(a_{j+N}) = f(a_j), \qquad j \in J, \tag{6.32}$$

and that the infinitesimal generator $\boldsymbol{\lambda}(x)$ for the process $\mathscr{M}(x)$ satisfies the conditions

$$\lambda_{ij}(x) = \lambda_{i-j}(x), \qquad i, j \in J, \tag{6.33}$$

and

$$\lambda_{ij}(x) \equiv 0, \qquad j < i \text{ if } J = Z^+. \tag{6.34}$$

It is not difficult to show that under the conditions (6.32)–(6.34) the process $f(\mathscr{M}(x))$ may be expressed as a function of a finite state Markov process, so that our results may be applied directly. In fact, define the process $M(x)$ by

$$M(x) = a_p \Leftrightarrow \mathscr{M}(x) = a_{p+jN}, \qquad j \in J\ (p = 1, \ldots, N). \tag{6.35}$$

Then, from (6.32),

$$f(M(x)) = f(\mathscr{M}(x)). \tag{6.36}$$

Moreover, it may be shown that the conditions (6.33) and (6.34) imply that

$M(x)$ is a Markov process and also that, for $0 \leqq x \leqq y$ and $p, q = 1, \ldots, N$,

$$\operatorname{Prob}\{M(y) = a_q | M(x) = a_p\} = \operatorname{Prob}\{M(y) = a_q | \mathcal{M}(x) = a_p\} \tag{6.37}$$
$$= \sum_{l \in J} \operatorname{Prob}\{\mathcal{M}(y) = a_{q+lN} | \mathcal{M}(x) = a_p\}.$$

It follows that the infinitesimal generator $\boldsymbol{\tau}(x)$ for the finite state Markov process $M(x)$ satisfies

$$\tau_{pq}(x) = \sum_{l \in J} \lambda_{p-q-lN}(x). \tag{6.38}$$

In Molyneux's models there is actually only one nonzero term in the sum in (6.38).

We now return to the system (6.1), but confine our attention to the case in which $M(x)$ has a stationary transition mechanism, so that $\boldsymbol{\tau}$ is independent of x. Then equations (6.11) and (6.20) have constant coefficients, and they may be solved by means of Laplace transforms. Thus, for a matrix function $\boldsymbol{F}(x)$, we define

$$\hat{\boldsymbol{F}}(s) = \mathcal{L}(\boldsymbol{F}) = \int_0^\infty e^{-sx} \boldsymbol{F}(x)\, dx. \tag{6.39}$$

First we consider the calculation of the first-order moments. Then, from (6.11), using the initial conditions in (6.12),

$$\boldsymbol{S}\hat{\boldsymbol{U}}_1 - \hat{\boldsymbol{V}}_1 = \boldsymbol{\alpha}^t, \qquad \boldsymbol{B}\hat{\boldsymbol{U}}_1 + \boldsymbol{S}\hat{\boldsymbol{V}}_1 = \boldsymbol{0}, \tag{6.40}$$

$$\boldsymbol{S}\hat{\boldsymbol{U}}_2 - \hat{\boldsymbol{V}}_2 = \boldsymbol{0}, \qquad \boldsymbol{B}\hat{\boldsymbol{U}}_2 + \boldsymbol{S}\hat{\boldsymbol{V}}_2 = \boldsymbol{\alpha}^t, \tag{6.41}$$

where

$$\boldsymbol{S} = s\boldsymbol{I}_N - \boldsymbol{\tau}^t. \tag{6.42}$$

These pairs of equations are readily solved to give

$$\hat{\boldsymbol{U}}_1(s) = (\boldsymbol{B} + \boldsymbol{S}^2)^{-1}\boldsymbol{S}\boldsymbol{\alpha}^t, \qquad \hat{\boldsymbol{V}}_1(s) = \boldsymbol{S}(\boldsymbol{B} + \boldsymbol{S}^2)^{-1}\boldsymbol{S}\boldsymbol{\alpha}^t - \boldsymbol{\alpha}^t, \tag{6.43}$$

and

$$\hat{\boldsymbol{U}}_2(s) = (\boldsymbol{B} + \boldsymbol{S}^2)^{-1}\boldsymbol{\alpha}^t, \qquad \hat{\boldsymbol{V}}_2(s) = \boldsymbol{S}(\boldsymbol{B} + \boldsymbol{S}^2)^{-1}\boldsymbol{\alpha}^t. \tag{6.44}$$

From (6.10), the Laplace transforms of the first-order moments are given by

$$\mathcal{L}(\langle u_m \rangle) = \boldsymbol{E}_N \hat{\boldsymbol{U}}_m(s), \qquad \mathcal{L}(\langle v_m \rangle) = \boldsymbol{E}_N \hat{\boldsymbol{V}}_m(s) \qquad (m = 1, 2). \tag{6.45}$$

Note, from (6.42)–(6.44), that the expressions in (6.45) are rational functions of s with denominator

$$D(s) = \det(\boldsymbol{B} + \boldsymbol{S}^2). \tag{6.46}$$

Next we consider the second-order moments. From (6.20), using the initial conditions in (6.23) and (6.24), it follows that, for $m = 0, 1, 2$,

$$\boldsymbol{S}\hat{\boldsymbol{X}}_m - 2\hat{\boldsymbol{Y}}_m = \delta_{m1}\boldsymbol{\alpha}^t, \tag{6.47}$$

$$\boldsymbol{B}\hat{\boldsymbol{X}}_m + \boldsymbol{S}\hat{\boldsymbol{Y}}_m - \hat{\boldsymbol{Z}}_m = \tfrac{1}{2}\delta_{m0}\boldsymbol{\alpha}^t, \tag{6.48}$$

and

$$2\boldsymbol{B}\hat{\boldsymbol{Y}}_m + \boldsymbol{S}\hat{\boldsymbol{Z}}_m = \delta_{m2}\boldsymbol{\alpha}^t, \tag{6.49}$$

where $\boldsymbol{S}$ is given by (6.42). These equations have the formal solutions

$$\hat{\boldsymbol{X}}_0 = \tfrac{1}{2}\boldsymbol{J}\boldsymbol{S}\boldsymbol{\alpha}^t, \qquad \hat{\boldsymbol{Y}}_0 = \tfrac{1}{2}\boldsymbol{S}\hat{\boldsymbol{X}}_0, \qquad \hat{\boldsymbol{Z}}_0 = (\boldsymbol{B} + \tfrac{1}{2}\boldsymbol{S}^2)\hat{\boldsymbol{X}}_0 - \tfrac{1}{2}\boldsymbol{\alpha}^t, \tag{6.50}$$

$$\begin{gathered} \hat{\boldsymbol{X}}_1 = \boldsymbol{J}(\boldsymbol{B} + \tfrac{1}{2}\boldsymbol{S}^2)\boldsymbol{\alpha}^t, \qquad \hat{\boldsymbol{Y}}_1 = \tfrac{1}{2}(\boldsymbol{S}\hat{\boldsymbol{X}}_1 - \boldsymbol{\alpha}^t), \\ \hat{\boldsymbol{Z}}_1 = (\boldsymbol{B} + \tfrac{1}{2}\boldsymbol{S}^2)\hat{\boldsymbol{X}}_1 - \tfrac{1}{2}\boldsymbol{S}\boldsymbol{\alpha}^t, \end{gathered} \tag{6.51}$$

$$\hat{\boldsymbol{X}}_2 = \boldsymbol{J}\boldsymbol{\alpha}^t, \qquad \hat{\boldsymbol{Y}}_2 = \tfrac{1}{2}\boldsymbol{S}\hat{\boldsymbol{X}}_2, \qquad \hat{\boldsymbol{Z}}_2 = (\boldsymbol{B} + \tfrac{1}{2}\boldsymbol{S}^2)\hat{\boldsymbol{X}}_2, \tag{6.52}$$

where

$$\boldsymbol{J} = (\boldsymbol{B}\boldsymbol{S} + \boldsymbol{S}\boldsymbol{B} + \tfrac{1}{2}\boldsymbol{S}^3)^{-1}. \tag{6.53}$$

From (6.29)–(6.31) the Laplace transforms of the second-order moments are given by

$$\begin{gathered} \mathscr{L}(\langle u_m^2 \rangle) = \boldsymbol{E}_N\hat{\boldsymbol{X}}_m(s), \qquad \mathscr{L}(\langle v_m^2 \rangle) = \boldsymbol{E}_N\hat{\boldsymbol{Z}}_m(s), \\ \mathscr{L}(\langle u_m v_m \rangle) = \boldsymbol{E}_N\hat{\boldsymbol{Y}}_m(s) \qquad (m = 1, 2), \end{gathered} \tag{6.54}$$

$$\mathscr{L}(\langle u_1 u_2 \rangle) = \boldsymbol{E}_N\hat{\boldsymbol{X}}_0(s), \qquad \mathscr{L}(\langle v_1 v_2 \rangle) = \boldsymbol{E}_N\hat{\boldsymbol{Z}}_0(s), \tag{6.55}$$

and

$$\mathscr{L}(\langle u_1 v_2 \rangle + \langle u_2 v_1 \rangle) = 2\boldsymbol{E}_N\hat{\boldsymbol{Y}}_0(s). \tag{6.56}$$

Note, from (6.50)–(6.53), that the expressions in (6.54)–(6.56) are rational functions of s with denominator

$$\Delta(s) = \det(2\boldsymbol{B}\boldsymbol{S} + 2\boldsymbol{S}\boldsymbol{B} + \boldsymbol{S}^3). \tag{6.57}$$

From (6.26) and (6.42), we obtain

$$\boldsymbol{E}_N\boldsymbol{\alpha}^t = 1, \qquad \boldsymbol{E}_N\boldsymbol{S} = s\boldsymbol{E}_N. \tag{6.58}$$

It then follows from the above relations that

$$\mathscr{L}(\langle v_1 \rangle) = s\mathscr{L}(\langle u_1 \rangle) - 1, \qquad \mathscr{L}(\langle v_2 \rangle) = s\mathscr{L}(\langle u_2 \rangle), \tag{6.59}$$

and

(6.60) $$\mathscr{L}(\langle u_1 v_1\rangle) = \tfrac{1}{2}[s\mathscr{L}(\langle u_1^2\rangle) - 1],$$

(6.61) $$\mathscr{L}(\langle u_2 v_2\rangle) = \tfrac{1}{2}s\mathscr{L}(\langle u_2^2\rangle),$$

(6.62) $$\mathscr{L}(\langle v_1 u_2\rangle + \langle u_1 v_2\rangle) = s\mathscr{L}(\langle u_1 u_2\rangle).$$

These relations are a direct consequence of the first equation in (6.1) and the initial conditions (6.2). From (6.28) it follows that

(6.63) $$\mathscr{L}(\langle u_1 v_2\rangle - \langle v_1 u_2\rangle) = 1/s.$$

We also note, from (6.42),

(6.64) $$\boldsymbol{\alpha}\boldsymbol{\tau} = 0 \Rightarrow \boldsymbol{S}\boldsymbol{\alpha}^\tau = s\boldsymbol{\alpha}^\tau,$$

and remark that the condition $\boldsymbol{\alpha}\boldsymbol{\tau} = 0$, where $\boldsymbol{\tau}$ is constant, implies that the process $M(x)$ is strictly stationary. When (6.64) holds we also have the relations

(6.65) $$\mathscr{L}(\langle u_1\rangle) = \mathscr{L}(\langle v_2\rangle),$$

(6.66) $$\mathscr{L}(\langle u_1 u_2\rangle) = \tfrac{1}{2}s\mathscr{L}(\langle u_2^2\rangle),$$

(6.67) $$\mathscr{L}(\langle v_1 v_2\rangle) = \tfrac{1}{2}s\mathscr{L}(\langle v_2^2\rangle) - \tfrac{1}{2},$$

as may readily be verified. Thus, for $\boldsymbol{\alpha}\boldsymbol{\tau} = 0$,

(6.68) $$\langle u_1(x)\rangle = \langle v_2(x)\rangle,$$

and, in view of the initial conditions in (6.2),

(6.69) $$\langle u_1(x)u_2(x)\rangle = \frac{1}{2}\frac{d}{dx}\langle[u_2(x)]^2\rangle,$$

(6.70) $$\langle v_1(x)v_2(x)\rangle = \frac{1}{2}\frac{d}{dx}\langle[v_2(x)]^2\rangle.$$

7. **Some particular examples for the wave equation.** In this section we consider the system (6.1), subject to the initial conditions (6.2), for particular cases of $f(M(x))$. One of the simplest cases is

(7.1) $$f(\mu) = \mu, \qquad M(x) = T(x),$$

where $T(x)$ is the random telegraph process, which assumes only the values ± 1. Thus $N = 2$ and we take $a_1 = 1$, $a_2 = -1$. The vector $\boldsymbol{\alpha}$ of initial probabilities and the infinitesimal generator $\boldsymbol{\tau}$ are given by

(7.2) $$\boldsymbol{\alpha} = (\tfrac{1}{2}, \tfrac{1}{2}) = \tfrac{1}{2}\boldsymbol{E}_2, \qquad \boldsymbol{\tau} = \begin{pmatrix} -b & b \\ b & -b \end{pmatrix} = \boldsymbol{\tau}^t,$$

where b is a constant. Note that $\boldsymbol{\alpha}\boldsymbol{\tau} = 0$, so that the process is stationary. From (6.6), (6.42), (7.1) and (7.2), we have

$$(7.3)\qquad \boldsymbol{B} = \beta_0^2\begin{bmatrix}(1+\varepsilon) & 0\\ 0 & (1-\varepsilon)\end{bmatrix}, \qquad \boldsymbol{S} = \begin{bmatrix}(s+b) & -b\\ -b & (s+b)\end{bmatrix}.$$

The authors [**8**] calculated the first- and second-order moments in the case $f(M(x)) = T(x)$, by proceeding in essentially the same manner as in the general case considered in this paper. It should be noted that probability density functions for the solution processes $\{u_m(x)\}$, $\{v_m(x)\}$ $(m = 1, 2)$ conditioned on $T(x)$ were used in [**8**], but, since $T(x) = \pm 1$ with equal probability, there are simple relations between the conditioned probability density functions and the unconditioned ones considered here. We merely quote the results for the Laplace transforms of the first- and second-order moments, which also follow in a straightforward manner from the results of the previous section.

Thus, for the first-order moments,

$$(7.4)\qquad \mathscr{L}(\langle u_2\rangle) = [(s+2b)^2 + \beta_0^2]/D(s),$$

where

$$(7.5)\qquad D(s) = \{(s^2 + \beta_0^2)[(s+2b)^2 + \beta_0^2] - \varepsilon^2\beta_0^4\},$$

and the Laplace transforms of the remaining moments are given by (6.59) and (6.65), since (6.64) holds. Hence,

$$(7.6)\qquad \langle u_2(x)\rangle = \sum_{r=1}^{4} a_r \exp(\sigma_r x),$$

where the σ_r are the four roots of the equation $D(s) = 0$, and the a_r are the residues of $\mathscr{L}(\langle u_2\rangle)$ at $s = \sigma_r$. Also,

$$(7.7)\qquad \langle u_1(x)\rangle = \langle v_2(x)\rangle = \frac{d}{dx}\langle u_2(x)\rangle, \qquad \langle v_1(x)\rangle = \frac{d}{dx}\langle u_1(x)\rangle.$$

It may be shown that, for $0 < \varepsilon < 1$, all the σ_r have negative real parts, so that the first-order moments decay exponentially as $x \to \infty$.

For the second-order moments it is found that

$$(7.8)\qquad \mathscr{L}(\langle u_1^2\rangle) = (1/\Delta)\{(s+2b)(s^2 + 2\beta_0^2)[(s+2b)^2 + 4\beta_0^2] - 8\varepsilon^2\beta_0^4(s+b)\},$$

$$(7.9)\qquad \mathscr{L}(\langle v_1^2\rangle) = (2/\Delta)\beta_0^4(s+2b)\{[(s+2b)^2 + 4\beta_0^2] + \varepsilon^2[s(s+2b) - 4\beta_0^2]\},$$

$$(7.10)\qquad \mathscr{L}(\langle u_2^2\rangle) = (2/\Delta)(s+2b)[(s+2b)^2 + 4\beta_0^2],$$

$$(7.11)\qquad \mathscr{L}(\langle v_2^2\rangle) = \mathscr{L}(\langle u_1^2\rangle),$$

where

$$\Delta = \Delta(s) \equiv \{s(s + 2b)(s^2 + 4\beta_0^2)[(s + 2b)^2 + 4\beta_0^2] - 16\varepsilon^2\beta_0^4(s + b)^2\}. \tag{7.12}$$

The Laplace transforms of the remaining moments are given by (6.60)-(6.63), (6.66) and (6.67). It follows that each second-order moment has the form $\sum_{l=0}^{6} c_l \exp(s_l x)$, where $s_0 = 0$ and the remaining s_l are the roots of $\Delta(s) = 0$. The c_l are constants, and $c_0 \neq 0$ only for $\langle u_1(x)v_2(x)\rangle$ and $\langle v_1(x)u_2(x)\rangle$. For small enough ε one of the s_l has a positive real part, so that the second-order moments increase exponentially as $x \to \infty$.

The striking difference between the behavior of the first- and second-order moments as $x \to \infty$ can be explained on the basis of phase cancellation. For each choice of the sample function $T(x)$, the solutions will be oscillatory with either bounded or unbounded amplitude as $x \to \infty$. However, on taking the stochastic average of all these solutions, cancellations occur between oscillatory solutions which are out of phase, and so the average decays exponentially as $x \to \infty$. However, if these solutions are squared before averaging, the cancellations cannot take place, and so the averages of the squares of the solutions increase exponentially as $x \to \infty$.

The authors [**8**] also considered the correlation functions, and obtained a result corresponding to that in (5.24). We refer the reader to [**8**] for the details. Explicit expressions for the elements of

$$\mathscr{L}\left[\sum_{q=1}^{2} (-1)^q \langle \boldsymbol{W} \times \boldsymbol{W}\rangle_q\right] \tag{7.13}$$

were given in [**42**], and we caution the reader that since probability density functions conditioned on $T(x)$ were used, the elements of $\langle \boldsymbol{W} \times \boldsymbol{W}\rangle_q$ in the notation of this paper are equal to $\frac{1}{2}$ of those given in that paper. Since, from (5.8),

$$\langle \boldsymbol{W}(x) \times \boldsymbol{W}(x)\rangle = \sum_{q=1}^{2} \langle \boldsymbol{W}(x) \times \boldsymbol{W}(x)\rangle_q, \tag{7.14}$$

and we have given the expressions for the elements of $\mathscr{L}(\langle \boldsymbol{W} \times \boldsymbol{W}\rangle)$, the expressions for the elements of the quantity in (7.13) suffice to determine those of $\mathscr{L}(\langle \boldsymbol{W} \times \boldsymbol{W}\rangle_q)$, $q = 1, 2$. It was found that, for $\zeta \geqq 0$, the correlation functions have the form

$$\langle \boldsymbol{W}(\xi + \zeta) \times \boldsymbol{W}(\xi)\rangle = \sum_{q=1}^{2} \sum_{r=1}^{4} \sum_{l=0}^{6} (\boldsymbol{B}_{r,q} \times \boldsymbol{I}_2)\boldsymbol{D}_{l,q} \exp(\sigma_r\zeta + s_l\xi), \tag{7.15}$$

where $\boldsymbol{B}_{r,q}$ are constant 2×2 matrices, the $\boldsymbol{D}_{l,q}$ are constant 4×4 matrices and, as before, the σ_r are the roots of $D(s) = 0$, where $D(s)$ is given by (7.5), the s_l $(l \neq 0)$ are the roots of $\Delta(s) = 0$, where $\Delta(s)$ is given by (7.12), and $s_0 = 0$. It is seen from (7.15) that the process $\boldsymbol{W}(x)$ is neither stationary nor wide-sense stationary.

The authors [**9**] calculated the first- and second-order moments of the solutions of (6.1) in the case in which $\boldsymbol{M}(x)$ is the vector process

$$\boldsymbol{M}(x) = (T_1(x), T_2(x)) \tag{7.16}$$

where $T_1(x)$ and $T_2(x)$ are stochastically independent random telegraph processes with rates b_1 and b_2, respectively, and

$$f[(\mu_1, \mu_2)] = c_1\mu_1 + c_2\mu_2, \tag{7.17}$$

where c_1 and c_2 are real numbers, with $c_1^2 + c_2^2 \neq 0$. From (7.16) and (7.17) it follows that

$$f(\boldsymbol{M}(x)) = c_1 T_1(x) + c_2 T_2(x). \tag{7.18}$$

The process $\boldsymbol{M}(x)$ has a state space consisting of the four vectors

$$\begin{aligned} \boldsymbol{a}_1 &= (1, 1), & \boldsymbol{a}_2 &= (1, -1), \\ \boldsymbol{a}_3 &= (-1, 1), & \boldsymbol{a}_4 &= (1, 1), \end{aligned} \tag{7.19}$$

and the initial probability vector

$$\boldsymbol{\alpha} = (\tfrac{1}{4}, \tfrac{1}{4}, \tfrac{1}{4}, \tfrac{1}{4}). \tag{7.20}$$

Since $T_1(x)$ and $T_2(x)$ are stochastically independent and Markovian, it is easy to see that $\boldsymbol{M}(x)$ is a finite state Markov chain and that, for $0 \leqq x \leqq y$,

$$\begin{aligned} \operatorname{Prob}\{\boldsymbol{M}(y) &= ((-1)^i, (-1)^j)\boldsymbol{M}(x) = ((-1)^k, (-1)^l)\} \\ &= \operatorname{Prob}\{T_1(y) = (-1)^i | T_1(x) = (-1)^k\} \\ &\quad \times \operatorname{Prob}\{T_2(y) = (-1)^j | T_2(x) = (-1)^l\}. \end{aligned} \tag{7.21}$$

From (7.19) and (7.21) it follows that the infinitesimal generator $\boldsymbol{\tau}$ for the process $\boldsymbol{M}(x)$ is given by

$$\boldsymbol{\tau} = (\boldsymbol{\tau}_1 \times \boldsymbol{I}_2) + (\boldsymbol{I}_2 \times \boldsymbol{\tau}_2), \tag{7.22}$$

where

$$\boldsymbol{\tau}_j = \begin{pmatrix} -b_j & b_j \\ b_j & -b_j \end{pmatrix}, \qquad j = 1, 2, \tag{7.23}$$

are the infinitesimal generators for the processes $T_1(x)$ and $T_2(x)$, respectively. Since $\boldsymbol{\tau}$ is constant, and $\boldsymbol{\alpha\tau} = 0$, the process $\boldsymbol{M}(x)$ is stationary. It is of interest to note that if $c_1^2 = c_2^2$ there is a collapsing of states in (7.18), from four to three, and, as may be shown with the help of a theorem due to Burke and Rosenblatt [**43**], $f(M(x))$ is not a Markov process in this case unless $b_1 = b_2$ also.

We merely quote the results of the calculations of the first- and second-order moments when (7.18) holds, and refer the reader to [**9**] for the details. All the first-order moments can be obtained from $\langle u_2(x)\rangle$ since (7.7) holds, and

$$\langle u_2(x)\rangle = \sum_{j=1}^{8} l_j(\varepsilon)\exp[\sigma_j(\varepsilon)x], \tag{7.24}$$

where the $\sigma_j(\varepsilon)$ are the roots of an eighth-order polynomial. For $0 \leqq \varepsilon \ll 1$ these roots have the form

$$\sigma_1 = \sigma_2^* = i\beta_0\left[1 - \frac{\varepsilon^2\beta_0^2}{8}\sum_{j=1}^{2}\frac{c_j^2}{b_j(b_j + i\beta_0)}\right] + O(\varepsilon^4), \tag{7.25}$$

$$\sigma_3 = \sigma_4^* = -2b_1 - 2b_2 + i\beta_0 + O(\varepsilon^2), \tag{7.26}$$

$$\sigma_5 = \sigma_6^* = -2b_1 + i\beta_0 + O(\varepsilon^2), \tag{7.27}$$

$$\sigma_7 = \sigma_8^* = -2b_2 + i\beta_0 + O(\varepsilon^2), \tag{7.28}$$

and the coefficients $l_j(\varepsilon)$ have the form

$$l_1(\varepsilon) = l_2^*(\varepsilon) = 1/(2i\beta_0) + O(\varepsilon^2), \tag{7.29}$$

$$l_j(\varepsilon) = l_{j+1}^*(\varepsilon) = O(\varepsilon^2), \qquad j = 3, 5, 7. \tag{7.30}$$

In the special case $c_1^2 = c_2^2$, $b_1 = b = b_2$, two of the roots are $(-2b \pm i\beta_0)$, and the corresponding coefficients in (7.24) are zero.

The second-order moments can all be obtained from $\langle v_1^2\rangle$, $\langle u_2^2\rangle$ and $\langle v_2^2\rangle = \langle u_1^2\rangle$, in view of the first equation in (6.1), the relation (6.28), and (6.69) and (6.70), which hold since $\boldsymbol{\alpha\tau} = 0$. It was found that

$$\langle[v_1(x)]^2\rangle = \sum_{k=1}^{12} m_k(\varepsilon)\exp[s_k(\varepsilon)x], \tag{7.31}$$

$$\langle[u_2(x)]^2\rangle = \sum_{k=1}^{12} n_k(\varepsilon)\exp[s_k(\varepsilon)x], \tag{7.32}$$

$$\langle[v_2(x)]^2\rangle = \sum_{k=1}^{12} p_k(\varepsilon)\exp[s_k(\varepsilon)x], \tag{7.33}$$

where the $s_k(\varepsilon)$ are the roots of a twelfth-order polynomial. For $0 \leqq \varepsilon \ll 1$ these roots have the form

$$s_1 = \frac{\varepsilon^2\beta_0^2}{2}\sum_{j=1}^{2}\frac{c_j^2 b_j}{(b_j^2+\beta_0^2)} + O(\varepsilon^4), \tag{7.34}$$

$$s_2 = s_3^* = 2i\beta_0 - \frac{\varepsilon^2\beta_0^2}{4}\sum_{j=1}^{2}\frac{c_j^2(b_j+2i\beta_0)}{b_j(b_j+i\beta_0)} + O(\varepsilon^4), \tag{7.35}$$

$$s_4 = -2b_1 - 2b_2 + O(\varepsilon^2), \tag{7.36}$$

$$s_5 = -2b_1 + O(\varepsilon^2), \qquad s_6 = -2b_2 + O(\varepsilon^2), \tag{7.37}$$

$$s_7 = s_8^* = -2b_1 - 2b_2 + 2i\beta_0 + O(\varepsilon^2), \tag{7.38}$$

$$s_9 = s_{10}^* = -2b_1 + 2i\beta_0 + O(\varepsilon^2), \tag{7.39}$$

$$s_{11} = s_{12}^* = -2b_2 + 2i\beta_0 + O(\varepsilon^2), \tag{7.40}$$

and the coefficients in (7.31)-(7.33) have the form

$$m_1(\varepsilon) = \tfrac{1}{2}\beta_0^2 + O(\varepsilon^2), \tag{7.41}$$

$$m_2(\varepsilon) = m_3^*(\varepsilon) = -\tfrac{1}{4}\beta_0^2 + O(\varepsilon^2), \tag{7.42}$$

$$n_1(\varepsilon) = 1/(2\beta_0^2) + O(\varepsilon^2), \tag{7.43}$$

$$n_2(\varepsilon) = n_3^*(\varepsilon) = -1/(4\beta_0^2) + O(\varepsilon^2), \tag{7.44}$$

$$p_1(\varepsilon) = \tfrac{1}{2} + O(\varepsilon^2), \tag{7.45}$$

$$p_2(\varepsilon) = p_3^*(\varepsilon) = \tfrac{1}{4} + O(\varepsilon^2), \tag{7.46}$$

and

$$m_k(\varepsilon) = O(\varepsilon^2), \qquad n_k(\varepsilon) = O(\varepsilon^2), \qquad p_k(\varepsilon) = O(\varepsilon^2), \qquad 4 \leqq k \leqq 12. \tag{7.47}$$

In the special case $c_1^2 = c_2^2$, $b_1 = b = b_2$, two of the roots are $(-2b \pm 2i\beta_0)$, there is a repeated root $-2b$, and only nine terms appear in the sums in (7.31)–(7.33).

Molyneux [7] investigated the first- and second-order moments for two particular cases of $f(M(x))$. Referring to the discussion in the previous section, in the first example, for which $J = Z^+$, the infinitesimal generator τ for the corresponding finite state Markov process $M(x)$ takes the form

$$\tau_{pq} = \lambda[-\delta_{pq} + \delta_{p,q-1} + \delta_{pN}\delta_{q1}] \qquad (p, q = 1, \ldots, N), \tag{7.48}$$

where λ is constant, $N = 2K$ and

$$f(a_p) = (p - 1 - K/2) = -f(a_{K+p}) \qquad (p = 1, \ldots, K). \tag{7.49}$$

In the second example, for which $J = Z$, the corresponding quantities are

$$\tau_{pq} = \lambda[-2\delta_{pq} + \delta_{p,q+1} + \delta_{p,q-1} + \delta_{p1}\delta_{qN} + \delta_{pN}\delta_{q1}], \tag{7.50}$$

for $p, q = 1, \ldots, N$, where $N = 4K'$ and, relabelling the states,

$$f(a_p) = (K' + 1 - p) = -f(a_{2K'+p}) \qquad (p = 1, \ldots, 2K'). \tag{7.51}$$

Remarkably, Molyneux was able to investigate the characteristic values of the matrices on the right-hand sides of (6.11) and (6.20), in the case $0 < \varepsilon \ll 1$, for these two particular examples. These characteristic values determine the behavior of the first- and second-order moments, respectively. However, Molyneux's conclusion that in the second example the second-order moments decay exponentially is not valid, as may be seen from his equation (64), which implies that $s_{20}^{(0)} > 0$, so that there is a positive characteristic value.

8. **A perturbation method for linear stochastic equations.** So far we have considered only exact methods, but now we turn to approximate ones. In this section we begin by discussing a perturbation method, which was developed independently by J. B. Keller [**10**], [**11**] and by R. C. Bourret [**12**], [**13**], [**14**], for calculating the expectations of the solutions of linear stochastic equations which contain a small parameter. We then investigate the application of the method to the calculation of the second-order moments, and the correlation functions, of the solutions of stochastic ordinary differential equations. We do this by considering a specific equation of the second-order, but the technique can be applied to the calculation of second- and higher-order moments for ordinary linear differential equations of any order.

We give a formulation of the perturbation method in the context in which it is needed here. Thus, consider a linear stochastic equation for the matrix $\boldsymbol{F}$ of the form

$$(\boldsymbol{L}_0 + \varepsilon \boldsymbol{L}_1)\boldsymbol{F} = 0, \tag{8.1}$$

where $\boldsymbol{L}_0$ is a nonstochastic matrix differential operator, but $\boldsymbol{L}_1$ is a stochastic one, and ε is a small positive parameter. It is supposed that the stochastic average of $\boldsymbol{L}_1$ is zero, that is,

$$\langle \boldsymbol{L}_1 \rangle = 0. \tag{8.2}$$

We define the incoherent part of $\boldsymbol{F}$ as

$$\subset \boldsymbol{F} \supset = \boldsymbol{F} - \langle \boldsymbol{F} \rangle, \tag{8.3}$$

so that

$$\langle \subset \boldsymbol{F} \supset \rangle = 0. \tag{8.4}$$

If we substitute for $\boldsymbol{F}$ from (8.3) into (8.1), and take the stochastic average, we obtain

$$(8.5) \qquad L_0\langle F\rangle + \varepsilon\langle L_1 \subset F \supset\rangle = 0.$$

We next subtract (8.5) from (8.1) and obtain

$$(8.6) \qquad L_0 \subset F \supset + \varepsilon L_1\langle F\rangle + \varepsilon[L_1 \subset F \supset - \langle L_1 \subset F \supset\rangle] = 0.$$

So far no approximations have been made, but if we neglect terms of order ε^2 in (8.6) we obtain the first-order approximation

$$(8.7) \qquad L_0 \subset F \supset = -\varepsilon L_1\langle F\rangle.$$

If the inverse operator L_0^{-1} is applied to this equation, and the resulting expression for $\subset F \supset$ is substituted into (8.5), the equation for $\langle F\rangle$ as given by Keller [**11**] is obtained, with terms of order ε^3 being neglected. However, the above form is more convenient for our purposes.

We now consider the solutions of the equations

$$(8.8) \qquad \frac{du_m}{dx} = v_m, \qquad \frac{dv_m}{dx} = -\beta_0^2[1 + \varepsilon N(x)]u_m \qquad (m = 1, 2),$$

which satisfy the nonstochastic initial conditions

$$(8.9) \qquad u_1(0) = 1 = v_2(0), \qquad u_2(0) = 0 = v_1(0),$$

where β_0 is a positive constant and $N(x)$ is a real, wide sense stationary stochastic process with

$$(8.10) \qquad \langle N(x)\rangle = 0, \qquad \langle N(x)N(\xi)\rangle = \Gamma(x - \xi).$$

The system (8.8) may be written in matrix form, with

$$(8.11) \qquad W(x) = \begin{bmatrix} u_1(x) & u_2(x) \\ v_1(x) & v_2(x) \end{bmatrix}.$$

If we define

$$A = \begin{bmatrix} 0 & -1 \\ \beta_0^2 & 0 \end{bmatrix}, \qquad C = \begin{bmatrix} 0 & 0 \\ 1 & 0 \end{bmatrix},$$

then, from (8.8),

$$(8.12) \qquad dW/dx + [A + \varepsilon\beta_0^2 N(x)C]W = 0,$$

with initial condition, from (8.9),

$$(8.13) \qquad W(0) = I,$$

where $I = I_2$ is the unit matrix of order 2. Equation (8.12) has the form (8.1), with $F = W$ and

$$(8.14) \qquad L_0 = I(d/dx) + A, \qquad L_1 = \beta_0^2 N(x)C.$$

Now let

$$\boldsymbol{P}(x) = \begin{bmatrix} \cos \beta_0 x & (1/\beta_0) \sin \beta_0 x \\ -\beta_0 \sin \beta_0 x & \cos \beta_0 x \end{bmatrix}. \tag{8.15}$$

It may be verified that

$$d\boldsymbol{P}/dx + \boldsymbol{A}\boldsymbol{P}(x) = 0, \qquad \boldsymbol{P}(0) = \boldsymbol{I}, \tag{8.16}$$

so that $\boldsymbol{P}$ is a fundamental matrix. Applying the first-order perturbation method, it follows from (8.7) that

$$\subset \boldsymbol{W}(x) \supset \; = -\varepsilon \beta_0^2 \int_0^x \boldsymbol{P}(x-\xi) N(\xi) \boldsymbol{C} \langle \boldsymbol{W}(\xi) \rangle \, d\xi, \tag{8.17}$$

and, from (8.5) and (8.10), that

$$\begin{aligned} &d\langle \boldsymbol{W}(x) \rangle / dx + \boldsymbol{A} \langle \boldsymbol{W}(x) \rangle \\ &\qquad = \varepsilon^2 \beta_0^4 \int_0^x \Gamma(x-\xi) \boldsymbol{C}\boldsymbol{P}(x-\xi) \boldsymbol{C} \langle \boldsymbol{W}(\xi) \rangle \, d\xi. \end{aligned} \tag{8.18}$$

From (8.13) the initial condition is $\langle \boldsymbol{W}(0) \rangle = \boldsymbol{I}$. But equation (8.18) may be solved by means of Laplace transforms, since the integral is of convolution type. Thus,

$$[s\boldsymbol{I} + \boldsymbol{A} - \varepsilon^2 \beta_0^4 \mathscr{L}(\Gamma \boldsymbol{C}\boldsymbol{P}\boldsymbol{C})] \mathscr{L}(\langle \boldsymbol{W} \rangle) = \boldsymbol{I}. \tag{8.19}$$

We define

$$\gamma(s) = \mathscr{L}(\Gamma) = \int_0^\infty e^{-sx} \Gamma(x) \, dx. \tag{8.20}$$

Then, as we found in [**44**], in component form

$$\mathscr{L}(\langle u_2 \rangle) = \{s^2 + \beta_0^2 - (\varepsilon^2 \beta_0^3 / 2i)[\gamma(s - i\beta_0) - \gamma(s + i\beta_0)]\}^{-1}, \tag{8.21}$$

and

$$\begin{aligned} \mathscr{L}(\langle u_1 \rangle) &= \mathscr{L}(\langle v_2 \rangle) = s\mathscr{L}(\langle u_2 \rangle), \\ \mathscr{L}(\langle v_1 \rangle) &= s\mathscr{L}(\langle u_1 \rangle) - 1. \end{aligned} \tag{8.22}$$

Consider first the case when $N(x) = D_0^{1/2} \zeta(x)$, where $\zeta(x)$ is Gaussian white noise, with zero mean and spectral density 1. Then $\Gamma(x) = D_0 \delta(x)$ and $\gamma(s) \;\; \frac{1}{2} D_0$. The factor $\frac{1}{2}$ arises since the lower limit in (8.20) is zero. It follows from (8.21) that $\mathscr{L}(\langle u_2 \rangle) = (s^2 + \beta_0^2)^{-1}$, and hence that the first-order moments correspond to the solution of the unperturbed equation, with $\varepsilon = 0$. This is identical with the exact result obtained in §3. Next, consider the important special case

(8.23) $$\Gamma(x) = e^{-2b|x|}, \qquad \gamma(s) = (s + 2b)^{-1}.$$

It follows that the first-order perturbation method gives the exact results when $N(x) = T(x)$, the random telegraph process, as is seen from (7.4) and (7.5), since (8.22) also holds in that case. Actually, Bourret [**15**] showed quite generally that, for linear differential equations with a random telegraph coefficient, the first-order perturbation method gives the exact results for the first-order moments. His results are couched in quantum mechanical terms, and he used diagram techniques to establish that the contributions to the higher-order perturbations cancel in this case. Our results give a direct verification for the particular equation (8.12).

However, let us now consider the case when

(8.24) $$N(x) = c_1 T_1(x) + c_2 T_2(x),$$

where $T_1(x)$ and $T_2(x)$ are stochastically independent random telegraph processes with rates b_1 and b_2, respectively, and $c_1 \neq 0 \neq c_2$. Then

(8.25) $$\Gamma(x) = c_1^2 \exp(-2b_1|x|) + c_2^2 \exp(-2b_2|x|),$$

and

(8.26) $$\gamma(s) = \sum_{j=1}^{2} c_j^2/(s + 2b_j).$$

It follows from (7.24) and (8.21) that the first-order perturbation method does not give the exact results for the first-order moments, even in the case $c_1^2 = c_2^2, b_1 = b = b_2$.

The perturbation results are, however, consistent with the exact results for $0 < \varepsilon \ll 1$ and $0 \leqq \varepsilon^2 \beta_0 x \leqq O(1)$. Thus, in [**42**] the Laplace transforms in (8.21) and (8.22) were inverted approximately under the assumption that $\gamma(s)$, as defined by (8.20), is analytic for $\mathrm{Re}(s) \geqq -a$, where $a > 0$ is independent of ε. This implies that the correlation function $\Gamma(x)$ is exponentially small for large x. It was found that, for sufficiently small ε and $0 \leqq \varepsilon^2 \beta_0 x \leqq O(1)$,

(8.27) $$\langle u_1(x) \rangle = \langle v_2(x) \rangle \approx \tfrac{1}{2}(\exp(\sigma_1 x) + \exp(\sigma_2 x)),$$

and

(8.28) $$\langle v_1(x) \rangle \approx \tfrac{1}{2} i\beta_0 (\exp(\sigma_1 x) - \exp(\sigma_2 x)),$$

(8.29) $$\langle u_2(x) \rangle \approx (-i/2\beta_0)(\exp(\sigma_1 x) - \exp(\sigma_2 x)),$$

where

(8.30) $$\sigma_1 = \sigma_2^* \approx i\beta_0 + \tfrac{1}{4}\varepsilon^2 \beta_0^2 [\gamma(2i\beta_0) - \gamma(0)].$$

Note the consistency of (8.30) with (7.25), when $\gamma(s)$ is given by (8.26), and of (8.29) with (7.24), (7.29) and (7.30). For general $\Gamma(x)$ the results are consistent with the perturbation results obtained by Papanicolaou and Keller [**16**], using a two-variable procedure.

We consider next the calculation of the second-order moments. From (8.3), taking Kronecker products,

$$\boldsymbol{F} \times \boldsymbol{F} = [\langle \boldsymbol{F} \rangle + \subset \boldsymbol{F} \supset] \times [\langle \boldsymbol{F} \rangle + \subset \boldsymbol{F} \supset]. \tag{8.31}$$

Taking expectations, and using (8.4), we obtain

$$\langle \boldsymbol{F} \times \boldsymbol{F} \rangle = (\langle \boldsymbol{F} \rangle \times \langle \boldsymbol{F} \rangle) + \langle \subset \boldsymbol{F} \supset \times \subset \boldsymbol{F} \supset \rangle. \tag{8.32}$$

But, in the first approximation, from (8.7),

$$\subset \boldsymbol{F} \supset = -\varepsilon \boldsymbol{L}_0^{-1} \boldsymbol{L}_1 \langle \boldsymbol{F} \rangle. \tag{8.33}$$

Hence it would seem that the second-order moments could be calculated from the first-order moments, by substituting for $\subset \boldsymbol{F} \supset$ from (8.33) into (8.32), a procedure which has been widely used, in one guise or another, in the literature. However, considering the particular system (8.8), and the relation (8.17) which corresponds to (8.33), while $\langle \subset \boldsymbol{W} \supset \times \subset \boldsymbol{W} \supset \rangle$ is of order ε^2 for $\beta_0 x = O(1)$, this is no longer true for $\beta_0 x = O(1/\varepsilon^2)$, as may be verified directly, using the approximations to the first-order moments given by (8.27)–(8.30). Moreover, from the exact results for the random telegraph process, we know that the first-order moments decay exponentially, while the second-order moments grow exponentially. Thus the above procedure for calculating the second-order moments is not a correct one when $\varepsilon^2 \beta_0 x = O(1)$.

In the next section we apply a different procedure, which does in fact give the exact results for the second-order moments when $N(x) = T(x)$. We also consider the calculation of the correlation functions.

9. **The second-order moments and correlation functions.** We begin by considering the second-order moments of the solutions of the system (8.8) and (8.9). We proceed, as in [**44**], to derive the equation satisfied by $\boldsymbol{W} \times \boldsymbol{W}$, and then to apply the first-order perturbation method to that equation. Thus, it follows from (8.12) that

$$(\boldsymbol{L}_0 + \varepsilon \boldsymbol{L}_1)[\boldsymbol{W}(x) \times \boldsymbol{W}(x)] = 0, \tag{9.1}$$

where

$$\boldsymbol{L}_0 = (\boldsymbol{I} \times \boldsymbol{I})\, d/dx + [(\boldsymbol{A} \times \boldsymbol{I}) + (\boldsymbol{I} \times \boldsymbol{A})], \tag{9.2}$$

and

$$\boldsymbol{L}_1 = \beta_0^2 N(x)[(\boldsymbol{C} \times \boldsymbol{I}) + (\boldsymbol{I} \times \boldsymbol{C})]. \tag{9.3}$$

From (8.13) the initial condition is

$$W(0) \times W(0) = I \times I. \tag{9.4}$$

Then, from the first-order approximation (8.7), it is found that

$$\subset W(x) \times W(x) \supset \; = \; -\varepsilon\beta_0^2 \int_0^x G(x-\xi)N(\xi)\langle W(\xi) \times W(\xi)\rangle \, d\xi, \tag{9.5}$$

where

$$G(x) = [P(x) \times P(x)][(C \times I) + (I \times C)], \tag{9.6}$$

and $P(x)$ satisfies (8.16). Hence, from (8.5), (9.2), (9.3) and (9.5),

$$\begin{aligned} d\langle W \times W\rangle/dx &+ [(A \times I) + (I \times A)]\langle W(x) \times W(x)\rangle \\ &= \varepsilon^2\beta_0^4 \int_0^x \Gamma(x-\xi)[(C \times I) + (I \times C)] \\ &\qquad \cdot G(x-\xi)\langle W(\xi) \times W(\xi)\rangle \, d\xi, \end{aligned} \tag{9.7}$$

with initial condition, from (9.4),

$$\langle W(0) \times W(0)\rangle = I \times I. \tag{9.8}$$

In [**44**] the authors solved equation (9.7), subject to (9.8), by means of Laplace transforms, and we merely quote the results. It was found that

$$\begin{aligned} \tfrac{1}{2}\mathscr{L}(\langle u_2^2\rangle) = \Omega \equiv \{&s(s^2 + 4\beta_0^2) \\ &+ \tfrac{1}{2}\varepsilon^2\beta_0^2[(s + 2i\beta_0)^2\gamma(s - 2i\beta_0) \\ &+ (s - 2i\beta_0)^2\gamma(s + 2i\beta_0) - 2s^2\gamma(s)]\}^{-1}, \end{aligned} \tag{9.9}$$

$$\begin{aligned} \mathscr{L}(\langle u_1^2\rangle) = \Omega\{&(s^2 + 2\beta_0^2) \\ &+ \tfrac{1}{2}\varepsilon^2\beta_0^2[(s + 2i\beta_0)\gamma(s - 2i\beta_0) \\ &+ (s - 2i\beta_0)\gamma(s + 2i\beta_0) - 2s\gamma(s)]\}, \end{aligned} \tag{9.10}$$

and

$$\begin{aligned} \mathscr{L}(\langle v_1^2\rangle) &- 2\Omega\beta_0^4 \\ &= \varepsilon^2\Omega\beta_0^4[(s + 2i\beta_0)\gamma(s - 2i\beta_0) + (s - 2i\beta_0)\gamma(s + 2i\beta_0)] \\ &+ \varepsilon^2\Omega\beta_0^6\{2\gamma(s - 2i\beta_0)\gamma(s + 2i\beta_0) \\ &- \gamma(s)[\gamma(s - 2i\beta_0) + \gamma(s + 2i\beta_0)]\}. \end{aligned} \tag{9.11}$$

Also, the Laplace transforms of the remaining second-order moments are given by (6.60)–(6.63), (6.66), (6.67) and (7.11).

It follows, from (7.8)–(7.10) and (7.12), that this application of the first-order perturbation method to the calculation of the second-order moments gives the exact results when $N(x) = T(x)$, the random telegraph process, so that (8.23) holds. Also, setting $\gamma(s) \equiv \frac{1}{2}D_0$, it follows from (3.10), (3.18) and (3.19) that the exact results are obtained when $N(x)$ is white noise. However, from (7.31)–(7.33) and (8.26), it follows that the exact results are not obtained for the second-order moments when $N(x)$ is the linear combination of two stochastically independent random telegraph processes, as given by (8.24), even in the case $c_1^2 = c_2^2, b_1 = b = b_2$.

The perturbation results are, however, consistent with the exact results. Thus, in [**42**] the Laplace transforms in (9.9)–(9.11) were inverted approximately under the assumption that $\gamma(s)$, as defined by (8.20), is analytic for $\mathrm{Re}(s) \geqq -a$, where $a > 0$ is independent of ε. For sufficiently small ε the expression for Ω in (9.9) has simple poles in the neighborhood of $s = 0$ and $s = \pm 2i\beta_0$, which we denote by s_1, s_2 and s_3. Their approximate values are given by

$$s_1 \approx \tfrac{1}{2}\varepsilon^2\beta_0^2[\gamma(2i\beta_0) + \gamma(-2i\beta_0)], \tag{9.12}$$

and

$$s_2 = s_3^* \approx 2i\beta_0 + \tfrac{1}{2}\varepsilon^2\beta_0^2[\gamma(2i\beta_0) - 2\gamma(0)]. \tag{9.13}$$

Note, from (8.10), (8.20) and (9.12), that

$$s_1 \approx \tfrac{1}{2}\varepsilon^2\beta_0^2 \int_{-\infty}^{\infty} \exp(2i\beta_0 x)\Gamma(x)\,dx \geqq 0, \tag{9.14}$$

by a well-known result [**45**]. For sufficiently small ε and $0 \leqq \varepsilon^2\beta_0 x \leqq O(1)$ it was found that

$$\langle[u_1(x)]^2\rangle = \langle[v_2(x)]^2\rangle \approx \tfrac{1}{4}(\exp(s_2x) + 2\exp(s_1x) + \exp(s_3x)), \tag{9.15}$$

$$\begin{aligned}\langle[v_1(x)]^2\rangle &\approx -\tfrac{1}{4}\beta_0^2(\exp(s_2x) - 2\exp(s_1x) + \exp(s_3x))\\ &\approx \beta_0^4\langle[u_2(x)]^2\rangle,\end{aligned} \tag{9.16}$$

$$\langle[u_1(x)u_2(x)]\rangle = \langle[u_2(x)v_2(x)]\rangle \approx \frac{-i}{4\beta_0}(\exp(s_2x) - \exp(s_3x)), \tag{9.17}$$

$$\langle[u_1(x)v_1(x)]\rangle = \langle[v_1(x)v_2(x)]\rangle \approx \tfrac{1}{4}i\beta_0(\exp(s_2x) - \exp(s_3x)), \tag{9.18}$$

and

$$\langle[u_1(x)v_2(x)]\rangle \approx \tfrac{1}{4}(\exp(s_2x) + 2 + \exp(s_3x)), \tag{9.19}$$

$$\langle[v_1(x)u_2(x)]\rangle \approx \tfrac{1}{4}(\exp(s_2x) - 2 + \exp(s_3x)). \tag{9.20}$$

The consistency of the above results with those in (7.31)–(7.47), when $\gamma(s)$

is given by (8.26), should be noted. For general $\Gamma(x)$, the results are consistent with those obtained by Papanicolaou and Keller [**16**], who used a two-variable procedure.

We now turn our attention to the calculation of the correlation functions, and describe the procedure developed in [**42**]. Thus, we define

$$J(x, \xi) = W(x + \xi) \times W(\xi), \qquad x \geqq 0. \tag{9.21}$$

From (8.12) it follows that

$$(L_0 + \varepsilon L_1)J(x, \xi) = 0, \tag{9.22}$$

where

$$L_0 = (I\partial/\partial x + A) \times I, \qquad L_1 = \beta_0^2 N(x + \xi)(C \times I). \tag{9.23}$$

The initial condition is

$$J(0, \xi) = W(\xi) \times W(\xi). \tag{9.24}$$

In applying the first-order perturbation procedure, we need to know both the expected value, and the incoherent part, of the quantity in (9.24). We will make use of the results obtained for the second-order moments, and in addition use the approximation to the incoherent part given by (9.5). Corresponding to (8.7) it is found that

$$\begin{aligned} \subset J(x, \xi) \supset &= [P(x) \times I] \subset W(\xi) \times W(\xi) \supset \\ &\quad - \varepsilon \beta_0^2 \int_0^x [P(x - \zeta) \times I] N(\zeta + \xi)(C \times I) \langle J(\zeta, \xi)\rangle \, d\zeta, \end{aligned} \tag{9.25}$$

where $P(x)$ satisfies (8.16). Then, corresponding to (8.5), making use of (9.5) in (9.25), we obtain

$$\begin{aligned} &\partial\langle J(x, \xi)\rangle/\partial x + (A \times I)\langle J(x, \xi)\rangle \\ &\quad - \varepsilon^2 \beta_0^4 \int_0^x \Gamma(x - \zeta)\{[CP(x - \zeta)C] \times I\}\langle J(\zeta, \xi)\rangle \, d\zeta \\ &= \varepsilon^2 \beta_0^4 \{[CP(x)] \times I\} \int_0^\xi \Gamma(x + \xi - \zeta) G(\xi - \zeta) \langle W(\zeta) \times W(\zeta)\rangle \, d\zeta. \end{aligned} \tag{9.26}$$

From (9.24) the initial condition is

$$\langle J(0, \xi)\rangle = \langle W(\xi) \times W(\xi)\rangle. \tag{9.27}$$

We first remark that it was shown in [**42**] that the correlation functions are given exactly by (9.26) in the case that $N(x) = T(x)$, the random telegraph process, in (8.8). This was established by solving (9.26), in the case $\Gamma(x) = e^{-2b|x|}$, with the help of Laplace transforms, and by direct comparison with the exact results obtained in [**8**]. Since we will demonstrate, in the next section, that the applications of the first-order perturbation method, discussed in this section, to the calculation of the second-order moments and the correlation functions of the solutions of the matrix system (5.1) and (5.2) gives the exact results when $M(x) = T(x)$, we omit the details of the proof given in [**42**] for the particular system (8.8).

We now consider the case of a general correlation function $\Gamma(x)$ and take Laplace transforms with respect to x in (9.26). Using the initial condition (9.27), it follows that

$$\begin{aligned}\{[s\boldsymbol{I} + \boldsymbol{A} - \varepsilon^2\beta_0^4\mathscr{L}(\Gamma\boldsymbol{CPC})] \times \boldsymbol{I}\}\mathscr{L}(\langle\boldsymbol{J}\rangle)&\\ = \langle\boldsymbol{W}(\xi) \times \boldsymbol{W}(\xi)\rangle + \mathscr{L}(\boldsymbol{R}),&\end{aligned} \tag{9.28}$$

where $\boldsymbol{R}$ denotes the expression on the right-hand side of equation (9.26). In view of (8.19), the expression for $\mathscr{L}(\langle\boldsymbol{J}\rangle)$ may be inverted in the form

$$\begin{aligned}\langle\boldsymbol{J}(x, \xi)\rangle - [\langle\boldsymbol{W}(x)\rangle \times \boldsymbol{I}]\langle\boldsymbol{W}(\xi) \times \boldsymbol{W}(\xi)\rangle&\\ = \varepsilon^2\beta_0^4\int_0^x \{[\langle\boldsymbol{W}(x-\eta)\rangle\boldsymbol{CP}(\eta)] \times \boldsymbol{I}\}&\\ \cdot\int_0^\xi \Gamma(\eta + \xi - \zeta)\boldsymbol{G}(\xi - \zeta)\langle\boldsymbol{W}(\zeta) \times \boldsymbol{W}(\zeta)\rangle\, d\zeta\, d\eta,&\end{aligned} \tag{9.29}$$

for $x \geqq 0$. Let us consider the double integral in (9.29). From (8.15) and (9.6), $\boldsymbol{P}(\eta)$ and $\boldsymbol{G}(\xi - \zeta)$ are bounded. From the approximations to the first- and second-order moments, for sufficiently small ε, $\langle\boldsymbol{W}(x)\rangle$ and $\langle\boldsymbol{W}(\xi) \times \boldsymbol{W}(\xi)\rangle$ are bounded on fixed intervals $0 \leqq \varepsilon^2\beta_0 x \leqq X$, $0 \leqq \varepsilon^2\beta_0\xi \leqq \Xi$, where X and Ξ are $O(1)$. Moreover,

$$\int_0^x\int_0^\xi |\Gamma(\eta + \xi - \zeta)|\, d\zeta\, d\eta = \int_0^x\int_0^\xi |\Gamma(\eta + \zeta)|\, d\zeta\, d\eta, \tag{9.30}$$

which is clearly bounded for $x \geqq 0$ and $\xi \geqq 0$, under our previous assumption on $\gamma(s)$, which implies that $\Gamma(x)$ is exponentially small for large x. It follows that the right-hand side of equation (9.29) is $O(\varepsilon^2)$ for $0 \leqq \varepsilon^2\beta_0 x \leqq X$ and $0 \leqq \varepsilon^2\beta_0\xi \leqq \Xi$, so that, from (9.21),

$$\langle\boldsymbol{W}(x + \xi) \times \boldsymbol{W}(\xi)\rangle \approx [\langle\boldsymbol{W}(x)\rangle \times \boldsymbol{I}]\langle\boldsymbol{W}(\xi) \times \boldsymbol{W}(\xi)\rangle. \tag{9.31}$$

This approximate relationship, expressing the correlation functions in terms of the first- and second-order moments, has been verified by Papanicolaou and Keller [**16**], using a two-variable procedure.

It is of interest to compare our approach to the calculation of the second-order moments and the correlation functions, to the approach proposed by J. B. Keller [**46**] and others [**47**]. Thus, considering the linear stochastic equation (8.1), where now $\boldsymbol{L}_0$ is a general nonstochastic matrix operator and $\boldsymbol{L}_1$ is a stochastic one, Keller proposes considering the equation

$$\{[\boldsymbol{L}_0(x) + \varepsilon \boldsymbol{L}_1(x)] \times [\boldsymbol{L}_0(\xi) + \varepsilon \boldsymbol{L}_1(\xi)]\}[\boldsymbol{F}(x) \times \boldsymbol{F}(\xi)] = 0, \tag{9.32}$$

and applying the perturbation method to calculate the correlation functions. For the case in which $\boldsymbol{L}_0$ and $\boldsymbol{L}_1$ are ordinary differential operators, the first-order perturbation method leads to a partial differential, double integral equation, which is not very tractable even for the particular system (8.8). Once the correlation functions are found by this method, the second-order moments are obtained by setting $x = \xi$. This contrasts with our approach, in which the second-order moments are calculated before the correlation functions. A limitation to our approach, however, is that it is not applicable when $\boldsymbol{L}_0$ and $\boldsymbol{L}_1$ are partial differential operators, since, in general, it is then not possible to derive a linear equation satisfied by $[\boldsymbol{F}(x) \times \boldsymbol{F}(x)]$.

10. **Equations with a random telegraph coefficient.** In this section we give a direct proof of Bourret's result [**15**] that, for linear differential equations with a random telegraph coefficient, the first-order perturbation method gives the exact results for the first-order moments of the solutions. We also show that the same is true for the second-order moments and the correlation functions. We consider systems of the form (5.1) and (5.2), with $M(x) = T(x)$, the random telegraph process, so that

$$d\boldsymbol{W}/dx = \boldsymbol{A}(T(x), x)\boldsymbol{W}(x), \tag{10.1}$$

where $\boldsymbol{W}$ is an $n \times m$ matrix. Without loss of generality, since $T(x) = \pm 1$, we may write

$$\boldsymbol{A}(T(x), x) = \boldsymbol{A}_0(x) + \varepsilon T(x)\boldsymbol{A}_1(x), \tag{10.2}$$

where the positive parameter ε is introduced for notational convenience. Similarly, the initial condition in (5.2) may be written in the form

$$\boldsymbol{W}(0) = \boldsymbol{w}_0 + \varepsilon T(0)\boldsymbol{w}_1. \tag{10.3}$$

We now apply the perturbation method discussed in §8. Corresponding to (8.1), with $\boldsymbol{F} = \boldsymbol{W}$, we have

$$\boldsymbol{L}_0 = \boldsymbol{I}_n\, d/dx - \boldsymbol{A}_0(x), \qquad \boldsymbol{L}_1 = -T(x)\boldsymbol{A}_1(x), \tag{10.4}$$

where $\boldsymbol{I}_n$ is the unit matrix of order n. We introduce the fundamental matrix $\boldsymbol{\Phi}(x)$ satisfying

$$\boldsymbol{L}_0\boldsymbol{\Phi}(x) = 0, \qquad \boldsymbol{\Phi}(0) = \boldsymbol{I}_n. \tag{10.5}$$

Then, corresponding to (8.7), since $\subset W(0)\supset = \varepsilon T(0)w_1$ from (10.3), we have

$$\subset W(x)\supset = \varepsilon\Phi(x)\left[T(0)w_1 + \int_0^x \Phi^{-1}(\xi)T(\xi)A_1(\xi)\langle W(\xi)\rangle\, d\xi\right]. \tag{10.6}$$

Hence, corresponding to (8.5),

$$\frac{d}{dx}\langle W(x)\rangle - A_0(x)\langle W(x)\rangle$$
$$= \varepsilon^2 A_1(x)\Phi(x)e^{-2bx}\left[w_1 + \int_0^x \Phi^{-1}(\xi)e^{2b\xi}A_1(\xi)\langle W(\xi)\rangle\, d\xi\right]. \tag{10.7}$$

From (10.3) the initial condition is

$$\langle W(0)\rangle = w_0. \tag{10.8}$$

We will now show that this equation for $\langle W\rangle$, which arises from the first-order perturbation method, is exact.

The exact results are contained in §5. Thus, corresponding to (5.9), making use of (7.2), we have

$$d\langle W\rangle_1/dx + b(\langle W\rangle_1 - \langle W\rangle_2) = [A_0(x) + \varepsilon A_1(x)]\langle W\rangle_1, \tag{10.9}$$

and

$$d\langle W\rangle_2/dx + b(\langle W\rangle_2 - \langle W\rangle_1) = [A_0(x) - \varepsilon A_1(x)]\langle W\rangle_2. \tag{10.10}$$

From (5.2), (5.10) and (10.3), the initial conditions are

$$\langle W(0)\rangle_1 = \tfrac{1}{2}(w_0 + \varepsilon w_1), \qquad \langle W(0)\rangle_2 = \tfrac{1}{2}(w_0 - \varepsilon w_1). \tag{10.11}$$

Also, from (5.8),

$$\langle W(x)\rangle = \langle W(x)\rangle_1 + \langle W(x)\rangle_2. \tag{10.12}$$

Thus, adding and subtracting equations (10.9) and (10.10),

$$L_0\langle W\rangle = \varepsilon A_1(x)(\langle W\rangle_1 - \langle W\rangle_2), \tag{10.13}$$

and

$$(L_0 + 2bI_n)(\langle W\rangle_1 - \langle W\rangle_2) = \varepsilon A_1(x)\langle W\rangle, \tag{10.14}$$

where L_0 is given by (10.4). From (10.5), (10.11) and (10.14), it follows that

$$e^{2bx}(\langle W\rangle_1 - \langle W\rangle_2)$$
$$= \varepsilon\Phi(x)\left[w_1 + \int_0^x \Phi^{-1}(\xi)e^{2b\xi}A_1(\xi)\langle W(\xi)\rangle\, d\xi\right]. \tag{10.15}$$

It is evident that (10.13) and (10.15) are equivalent to (10.7), so that the result is established for the first-order moments [**53**].

Next we consider the second-order moments, and introduce the notations $\boldsymbol{W}^{(2)} = \boldsymbol{W} \times \boldsymbol{W}$ and

$$(\boldsymbol{B}, \boldsymbol{C}) = (\boldsymbol{B} \times \boldsymbol{C}) + (\boldsymbol{C} \times \boldsymbol{B}), \qquad [[A]] = (A, \boldsymbol{I}_n). \tag{10.16}$$

Then, from (10.1) and (10.2) it follows that

$$d\boldsymbol{W}^{(2)}/dx = \{[[\boldsymbol{A}_0(x)]] + \varepsilon T(x)[[\boldsymbol{A}_1(x)]]\}\boldsymbol{W}(x)^{(2)}. \tag{10.17}$$

Also, from (10.3), since $[T(0)]^2 = 1$, the initial condition is

$$\boldsymbol{W}(0)^{(2)} = \boldsymbol{w}_0^{(2)} + \varepsilon^2 \boldsymbol{w}_1^{(2)} + \varepsilon T(0)(\boldsymbol{w}_0, \boldsymbol{w}_1). \tag{10.18}$$

Since the equation for $\boldsymbol{W}^{(2)}$ has the same structure as that for $\boldsymbol{W}$, it follows that the first-order perturbation method gives the exact results for the second-order moments, provided that the full initial conditions in (10.18) are used. However, since the incoherent part of $\boldsymbol{W}^{(2)}$, as well as its expected value, is needed in applying the perturbation method to the calculation of the correlation functions, we give some of the pertinent formulas. In particular, corresponding to (8.7), it is found from (10.17) and (10.18) that

$$\begin{aligned} \subset \boldsymbol{W}(x)^{(2)} \supset = \varepsilon \boldsymbol{\Phi}(x)^{(2)} \Big\{ & T(0)(\boldsymbol{w}_0, \boldsymbol{w}_1) \\ & + \int_0^x \boldsymbol{\Phi}^{-1}(\xi)^{(2)} T(\xi)[[\boldsymbol{A}_1(\xi)]]\langle \boldsymbol{W}(\xi)^{(2)} \rangle \, d\xi \Big\}. \end{aligned} \tag{10.19}$$

Also, according to the exact results, from (5.8), (5.11), (7.2) and (10.17), we merely replace $\boldsymbol{W}$, $\boldsymbol{A}_0(x)$ and $\boldsymbol{A}_1(x)$ in (10.9), (10.10) and (10.12) by $\boldsymbol{W}^{(2)}$, $[[\boldsymbol{A}_0(x)]]$ and $[[\boldsymbol{A}_1(x)]]$ respectively. From (5.2), (5.12) and (10.3), the initial conditions are

$$\langle \boldsymbol{W}(0)^{(2)} \rangle_p = \tfrac{1}{2}[\boldsymbol{w}_0^{(2)} + \varepsilon^2 \boldsymbol{w}_1^{(2)} + (-1)^{p-1}\varepsilon(\boldsymbol{w}_0, \boldsymbol{w}_1)], \tag{10.20}$$

for $p = 1, 2$. Corresponding to (10.13) and (10.14) we have

$$\begin{aligned} d\langle \boldsymbol{W}^{(2)} \rangle/dx - [[\boldsymbol{A}_0(x)]]\langle \boldsymbol{W}^{(2)} \rangle \\ = \varepsilon[[\boldsymbol{A}_1(x)]](\langle \boldsymbol{W}^{(2)} \rangle_1 - \langle \boldsymbol{W}^{(2)} \rangle_2), \end{aligned} \tag{10.21}$$

and

$$\begin{aligned} & d(\langle \boldsymbol{W}^{(2)} \rangle_1 - \langle \boldsymbol{W}^{(2)} \rangle_2)/dx + 2b(\langle \boldsymbol{W}^{(2)} \rangle_1 - \langle \boldsymbol{W}^{(2)} \rangle_2) \\ & \qquad = [[\boldsymbol{A}_0(x)]](\langle \boldsymbol{W}^{(2)} \rangle_1 - \langle \boldsymbol{W}^{(2)} \rangle_2) + \varepsilon[[\boldsymbol{A}_1(x)]]\langle \boldsymbol{W}^{(2)} \rangle. \end{aligned} \tag{10.22}$$

From (10.4), (10.5), (10.16), (10.20) and (10.22) it follows that

$$e^{2bx}(\langle W^{(2)}\rangle_1 - \langle W^{(2)}\rangle_2)$$

$$(10.23)\qquad = \varepsilon\mathbf{\Phi}(x)^{(2)}\left\{(w_0, w_1) + \int_0^x \mathbf{\Phi}^{-1}(\xi)^{(2)}e^{2b\xi}[[A_1(\xi)]]\langle W(\xi)^{(2)}\rangle\, d\xi\right\}.$$

Equations (10.21) and (10.23) lead to the equation for $\langle W^{(2)}\rangle$ which corresponds to (8.5), with $\subset W^{(2)}\supset$ given by (10.19). The initial condition is, from (10.18) or (10.20),

$$(10.24)\qquad \langle W(0)^{(2)}\rangle = w_0^{(2)} + \varepsilon^2 w_1^{(2)}.$$

We now turn to the calculation of the correlation functions by the first-order perturbation method. We define

$$(10.25)\qquad K(x, \xi) = W(x) \times W(\xi), \qquad 0 \leqq \xi \leqq x.$$

From (10.1) and (10.2) it follows that

$$(10.26)\qquad \partial K/\partial x = \{[A_0(x) + \varepsilon T(x)A_1(x)] \times I_n\}K(x, \xi).$$

The initial condition is, from (10.25),

$$(10.27)\qquad K(\xi, \xi) = W(\xi)^{(2)}.$$

Corresponding to (8.7) it is found that

$$\subset K(x, \xi)\supset = [\mathbf{\Phi}(x) \times I_n]\Big\{[\mathbf{\Phi}^{-1}(\xi) \times I_n]\subset W(\xi)^{(2)}\supset$$

$$(10.28)\qquad + \int_\xi^x T(\eta)\{[\mathbf{\Phi}^{-1}(\eta)A_1(\eta)] \times I_n\}\langle K(\eta, \xi)\rangle\, d\eta\Big\}.$$

We make use of the expression for $\subset W^{(2)}\supset$ given in (10.19). Then, corresponding to (8.5), it is found, using (10.23), that

$$\partial\langle K(x, \xi)\rangle/\partial x - [A_0(x) \times I_n]\langle K(x, \xi)\rangle$$

$$(10.29)\qquad -\varepsilon^2\int_\xi^x e^{-2b(x-\eta)}\{[A_1(x)\mathbf{\Phi}(x)\mathbf{\Phi}^{-1}(\eta)A_1(\eta)] \times I_n\}\langle K(\eta, \xi)\rangle\, d\eta$$

$$= \varepsilon e^{-2b(x-\xi)}\{[A_1(x)\mathbf{\Phi}(x)\mathbf{\Phi}^{-1}(\xi)] \times I_n\}(\langle W(\xi)^{(2)}\rangle_1 - \langle W(\xi)^{(2)}\rangle_2).$$

From (10.27) the initial condition is

$$(10.30)\qquad \langle K(\xi, \xi)\rangle = \langle W(\xi)^{(2)}\rangle = \langle W(\xi)^{(2)}\rangle_1 + \langle W(\xi)^{(2)}\rangle_2,$$

from (5.8).

The exact expression for the correlation functions follows from the results of §5. From (5.24) we have, for $0 \leqq \xi \leqq x$,

(10.31) $$\langle \boldsymbol{W}(x) \times \boldsymbol{W}(\xi)\rangle = \sum_{q=1}^{2} \left[\sum_{p=1}^{2} \boldsymbol{\Psi}_{pq}(x, \xi) \times \boldsymbol{I}_n \right] \langle \boldsymbol{W}(\xi)^{(2)}\rangle_q,$$

where, from (5.19), (7.2) and (10.2),

(10.32) $$\partial \boldsymbol{\Psi}_{1q}/\partial x + b(\boldsymbol{\Psi}_{1q} - \boldsymbol{\Psi}_{2q}) = [\boldsymbol{A}_0(x) + \varepsilon \boldsymbol{A}_1(x)]\boldsymbol{\Psi}_{1q},$$

and

(10.33) $$\partial \boldsymbol{\Psi}_{2q}/\partial x + b(\boldsymbol{\Psi}_{2q} - \boldsymbol{\Psi}_{1q}) = [\boldsymbol{A}_0(x) - \varepsilon \boldsymbol{A}_1(x)]\boldsymbol{\Psi}_{2q}.$$

The initial conditions are, from (5.20),

(10.34) $$\boldsymbol{\Psi}_{pq}(\xi, \xi) = \delta_{pq}\boldsymbol{I}_n.$$

Adding and subtracting equations (10.32) and (10.33),

(10.35) $$\boldsymbol{L}_0(\boldsymbol{\Psi}_{1q} + \boldsymbol{\Psi}_{2q}) = \varepsilon \boldsymbol{A}_1(x)(\boldsymbol{\Psi}_{1q} - \boldsymbol{\Psi}_{2q}),$$

and

(10.36) $$(\boldsymbol{L}_0 + 2b\boldsymbol{I}_n)(\boldsymbol{\Psi}_{1q} - \boldsymbol{\Psi}_{2q}) = \varepsilon \boldsymbol{A}_1(x)(\boldsymbol{\Psi}_{1q} + \boldsymbol{\Psi}_{2q}),$$

where $\boldsymbol{L}_0$ is given by (10.4).

Let us define the integrodifferential operator $\boldsymbol{Q}$ by

(10.37) $$\boldsymbol{Q}[\boldsymbol{F}(x)] = \partial \boldsymbol{F}/\partial x - \boldsymbol{A}_0(x)\boldsymbol{F}(x) - \varepsilon^2 \int_{\xi}^{x} e^{-2b(x-\eta)}\boldsymbol{A}_1(x)\boldsymbol{\Phi}(x)\boldsymbol{\Phi}^{-1}(\eta)\boldsymbol{A}_1(\eta)\boldsymbol{F}(\eta)\, d\eta.$$

Then, solving (10.36) for $(\boldsymbol{\Psi}_{1q} - \boldsymbol{\Psi}_{2q})$, using (10.5) and (10.34), and substituting into (10.35), it is found that

(10.38) $$\boldsymbol{Q}\left[\sum_{p=1}^{2} \boldsymbol{\Psi}_{pq}(x, \xi) \right] = \varepsilon(\delta_{1q} - \delta_{2q})e^{-2b(x-\xi)}\boldsymbol{A}_1(x)\boldsymbol{\Phi}(x)\boldsymbol{\Phi}^{-1}(\xi).$$

Thus,

(10.39) $$\boldsymbol{Q}\left[\sum_{p=1}^{2} \sum_{q=1}^{2} \boldsymbol{\Psi}_{pq}(x, \xi) \right] = 0,$$

and

(10.40) $$\boldsymbol{Q}\left[\sum_{p=1}^{2} \sum_{q=1}^{2} (-1)^q \boldsymbol{\Psi}_{pq}(x, \xi) \right] = -2\varepsilon e^{-2b(x-\xi)}\boldsymbol{A}_1(x)\boldsymbol{\Phi}(x)\boldsymbol{\Phi}^{-1}(\xi).$$

From (10.34) the initial conditions are

(10.41) $$\sum_{p=1}^{2} \sum_{q=1}^{2} \boldsymbol{\Psi}_{pq}(\xi, \xi) = 2\boldsymbol{I}_n, \qquad \sum_{p=1}^{2} \sum_{q=1}^{2} (-1)^q \boldsymbol{\Psi}_{pq}(\xi, \xi) = 0.$$

It follows, from (10.29), (10.30), (10.37) and (10.39)–(10.41), that $\langle \boldsymbol{K}(x, \xi) \rangle$ is given by the expression on the right-hand side of equation (10.31). Thus, the first-order perturbation method gives the exact results for the correlation functions.

11. **Application of a limit theorem.** In this section we consider the application of a limit theorem of R. Z. Has'minskiĭ [**17**] to the investigation, for $0 < \varepsilon \ll 1$, of the solutions of the system (8.8)–(8.10). In particular we shall obtain a formulation for calculating expectations of functions of u_1, v_1, u_2 and v_2, on an interval $0 \leqq \varepsilon^2 \beta_0 x \leqq O(1)$. The limit theorem applies to differential equations with random right-hand sides of the form

$$dy_j/dx = \varepsilon F_j(\boldsymbol{y}, x) \qquad (j = 1, \ldots, n), \tag{11.1}$$

satisfying the nonstochastic initial condition $\boldsymbol{y}(0) = \boldsymbol{y}_0$. It is supposed that, for fixed $\boldsymbol{y}$,

$$\langle F_j(\boldsymbol{y}, x) \rangle \equiv 0 \qquad (j = 1, \ldots, n). \tag{11.2}$$

Now define, for fixed $\boldsymbol{y}$,

$$K_j(\boldsymbol{y}, z, x) = \langle \partial F_j(\boldsymbol{y}, z)/\partial y_k F_k(\boldsymbol{y}, x) \rangle, \tag{11.3}$$

where it is understood that repeated indices are summed, and

$$a_{jk}(\boldsymbol{y}, z, x) = \langle F_j(\boldsymbol{y}, z) F_k(\boldsymbol{y}, x) \rangle. \tag{11.4}$$

We consider the case in which

$$K_j(\boldsymbol{y}, z + \tau, x + \tau) = K_j(\boldsymbol{y}, z, x), \tag{11.5}$$

and

$$a_{jk}(\boldsymbol{y}, z + \tau, x + \tau) = a_{jk}(\boldsymbol{y}, z, x), \tag{11.6}$$

for some fixed τ. Then, Has'minskiĭ's definitions of the quantities $\overline{K}_j$ and $\bar{a}_{jk}$ become [**17**]

$$\overline{K}_j(\boldsymbol{y}) = \frac{1}{\tau} \int_0^\tau \int_0^\infty K_j(\boldsymbol{y}, z, z - \zeta)\, d\zeta\, dz, \tag{11.7}$$

and

$$\bar{a}_{jk}(\boldsymbol{y}) = \frac{1}{\tau} \int_0^\tau \int_{-\infty}^\infty a_{jk}(\boldsymbol{y}, z, x)\, dx\, dz. \tag{11.8}$$

Under the appropriate conditions, for the statement of which we refer the reader to Has'minskiĭ's paper [**17**], the limit theorem states that, on the interval $0 \leqq \xi < \xi_0$, where ξ_0 is an arbitrary positive number, the process $\boldsymbol{y}(\xi/\varepsilon^2)$ converges weakly as $\varepsilon \to 0$ to a Markov process $\boldsymbol{Y}(\xi)$, which is continuous with probability 1, and whose local properties are given by

$$E\{\Delta Y_j(\xi)|Y(\xi) = Y\} = \bar{K}_j(Y)\Delta\xi + o(\Delta\xi), \tag{11.9}$$

$$E\{\Delta Y_j(\xi)\Delta Y_k(\xi)|Y(\xi) = Y\} = \bar{a}_{jk}(Y)\Delta\xi + o(\Delta\xi). \tag{11.10}$$

As a simple application of this limit theorem we consider the scalar equation, dropping the subscript 1,

$$dy/dx = \varepsilon N(x), \qquad y(0) = 0, \tag{11.11}$$

where $N(x)$ is a bounded, zero mean, wide sense stationary process, with correlation function given by (8.10). The theorem is applicable if $N(x)$ satisfies a certain strong mixing condition [**17**]. Then, from (11.3), $K = 0$ and, from (8.10) and (11.4),

$$a(y, z, x) = \Gamma(|x - z|). \tag{11.12}$$

It follows from (11.5)–(11.8) that $\bar{K} = 0$, and

$$\bar{a} = 2\int_0^\infty \Gamma(x)\,dx. \tag{11.13}$$

Hence, from (11.9) and (11.10), the probability density function $P(Y, \xi)$ of the limit process satisfies

$$\partial P/\partial\xi = \tfrac{1}{2}\bar{a}\partial^2 P/\partial Y^2, \qquad P(Y, 0) = \delta(Y), \tag{11.14}$$

in view of the initial condition $y(0) = 0$. It follows that

$$P(Y, \xi) = (2\pi\bar{a}\xi)^{-1/2}\exp[-Y^2/2\bar{a}\xi]. \tag{11.15}$$

If we choose $\Gamma(x) = e^{-2b|x|}$, and hence $\bar{a} = 1/b$, with the aid of (11.15) we can obtain again the result described in equation (2.30).

Now let us consider the more complicated system (8.8) and (8.9). Has′minskiĭ [**17**] applied his results to just a single equation of this form. Following his procedure, we let

$$\begin{aligned} u_m &= \exp(w_m)\cos(\beta_0 x + \varphi_m), \\ v_m &= -\beta_0\exp(w_m)\sin(\beta_0 x + \varphi_m), \end{aligned} \tag{11.16}$$

for $m = 1, 2$. It follows that

$$dw_m/dx = \tfrac{1}{2}\varepsilon\beta_0 N(x)\sin 2(\beta_0 x + \varphi_m), \tag{11.17}$$

$$d\varphi_m/dx = \varepsilon\beta_0 N(x)\cos^2(\beta_0 x + \varphi_m). \tag{11.18}$$

Thus the vector

$$\mathbf{y} = (y_1, y_2, y_3, y_4) = (w_1, \varphi_1, w_2, \varphi_2) \tag{11.19}$$

satisfies an equation of the form (11.1), with $n = 4$. Moreover, if $N(x)$ satisfies (8.10), then (11.2) is satisfied, and so are (11.5) and (11.6), with $\tau = \pi/\beta_0$. The limit theorem is applicable if, in addition, $N(x)$ is bounded and satisfies a certain strong mixing condition [**17**]. The quantities $\bar{K}_j$ and $\bar{a}_{jk}$, defined by (11.7) and (11.8), were calculated in [**18**]. Let

$$a = \frac{\beta_0^2}{4}\int_0^\infty \Gamma(x)\cos(2\beta_0 x)\,dx, \qquad c = \frac{\beta_0^2}{2}\int_0^\infty \Gamma(x)\,dx, \tag{11.20}$$

and

$$\kappa = \frac{-\beta_0^2}{4}\int_0^\infty \Gamma(x)\sin(2\beta_0 x)\,dx. \tag{11.21}$$

Then, it was found that

$$\bar{K}_1 = a = \bar{K}_3, \qquad \bar{K}_2 = \kappa = \bar{K}_4, \tag{11.22}$$

$$\bar{a}_{11} = a = \bar{a}_{33}, \qquad \bar{a}_{22} = (c + a) = \bar{a}_{44}, \tag{11.23}$$

$$\bar{a}_{jk} = \bar{a}_{kj}, \qquad \bar{a}_{12} = 0 = \bar{a}_{34}, \tag{11.24}$$

$$\bar{a}_{13} = a\cos 2(\varphi_1 - \varphi_2), \qquad \bar{a}_{24} = [c + a\cos 2(\varphi_1 - \varphi_2)], \tag{11.25}$$

and

$$\bar{a}_{14} = a\sin 2(\varphi_1 - \varphi_2) = -\bar{a}_{23}. \tag{11.26}$$

Now consider the transition probability density $p(\xi, \boldsymbol{\eta}; \boldsymbol{Y})$ for the event $\boldsymbol{Y}(\xi + \xi_1)$, given $\boldsymbol{Y}(\xi_1) = \boldsymbol{Y}$, depending on the increment ξ, but not on ξ_1. From (11.9) and (11.10), p satisfies the backward equation [**1**]

$$\frac{\partial p}{\partial \xi} = \bar{K}_j(\boldsymbol{Y})\frac{\partial p}{\partial Y_j} + \frac{1}{2}\bar{a}_{jk}(\boldsymbol{Y})\frac{\partial^2 p}{\partial Y_j \partial Y_k}; \qquad p(0, \boldsymbol{\eta}; \boldsymbol{Y}) = \delta(\boldsymbol{Y} - \boldsymbol{\eta}). \tag{11.27}$$

In view of (11.16) and (11.19), since we are interested in calculating the expectation of a function of u_1, v_1, u_2 and v_2, we introduce the conditional moment

$$g(\xi, \boldsymbol{Y}, x) = \int G(\boldsymbol{\eta}, x + \xi/\varepsilon^2)p(\xi, \boldsymbol{\eta}; \boldsymbol{Y})\,d\boldsymbol{\eta}, \tag{11.28}$$

where $G(\boldsymbol{\eta}, \theta)$ is periodic in θ, with fixed period, the integration being over the full range of variables. Then, from (11.27) and (11.28), it follows that

$$\frac{\partial g}{\partial \xi} = \frac{1}{\varepsilon^2}\frac{\partial g}{\partial x} + \bar{K}_j(\boldsymbol{Y})\frac{\partial g}{\partial Y_j} + \frac{1}{2}\bar{a}_{jk}(\boldsymbol{Y})\frac{\partial^2 g}{\partial Y_j \partial Y_k}, \tag{11.29}$$

with initial condition

$$g(0, \boldsymbol{Y}, x) = G(\boldsymbol{Y}, x). \tag{11.30}$$

Moreover,

$$g(\xi, \boldsymbol{Y}(0), 0) = \langle G(\boldsymbol{Y}(\xi), \xi/\varepsilon^2)\rangle \approx \langle G(\boldsymbol{y}(x), x)\rangle, \tag{11.31}$$

for $0 < \varepsilon \ll 1$, where $\xi = \varepsilon^2 x$, the approximate relationship in (11.31) being meaningful under appropriate conditions on $G(\boldsymbol{Y}, \theta)$. Thus, if $G(\boldsymbol{Y}, \theta)$ is bounded, and continuous in $\boldsymbol{Y}$, the weak convergence of $\boldsymbol{y}(\xi/\varepsilon^2)$ to $\boldsymbol{Y}(\xi)$ implies [**48**] that $\langle G(\boldsymbol{Y}(\xi), \theta)\rangle - \langle G(\boldsymbol{y}(x), \theta)\rangle \to 0$, as $\varepsilon \to 0$, for fixed θ. Further, if $G(\boldsymbol{Y}, \theta)$, in addition to being periodic in θ, is continuous in θ uniformly in $\boldsymbol{Y}$, then it may be shown in a straightforward manner that the convergence to zero is uniform in θ. Under these circumstances it is meaningful to set $\theta = \xi/\varepsilon^2$, with ξ fixed.

For the particular problem under consideration, we now rewrite equation (11.29) in terms of the original variables (u_m, v_m), which we capitalize to indicate that the limit process is under consideration. After some tedious calculations it is found, from (11.16), (11.19) and (11.22)–(11.26), that, with $g = q$,

$$\begin{aligned}\frac{\partial q}{\partial \xi} = {}& (\kappa + \beta_0/\varepsilon^2)\left[\frac{1}{\beta_0}\left(V_1\frac{\partial q}{\partial U_1} + V_2\frac{\partial q}{\partial U_2}\right) - \beta_0\left(U_1\frac{\partial q}{\partial V_1} + U_2\frac{\partial q}{\partial V_2}\right)\right] \\ & + \frac{1}{2}(c + a)\left[\frac{1}{\beta_0}\left(V_1\frac{\partial}{\partial U_1} + V_2\frac{\partial}{\partial U_2}\right) - \beta_0\left(U_1\frac{\partial}{\partial V_1} + U_2\frac{\partial}{\partial V_2}\right)\right]^2 q \\ & + \frac{a}{2}\left(U_1\frac{\partial}{\partial U_1} + V_1\frac{\partial}{\partial V_1} + U_2\frac{\partial}{\partial U_2} + V_2\frac{\partial}{\partial V_2} + 1\right)^2 q - \frac{a}{2}q \\ & + 2a(U_1V_2 - V_1U_2)\left(\frac{\partial^2 q}{\partial V_1 \partial U_2} - \frac{\partial^2 q}{\partial U_1 \partial V_2}\right),\end{aligned} \tag{11.32}$$

it being sufficient for our purposes to consider the case in which q does not depend explicitly on x. Corresponding to (11.30) and (11.31) we have

$$q(0, U_1, V_1, U_2, V_2) = Q(U_1, V_1, U_2, V_2), \tag{11.33}$$

and

$$\langle Q(U_1(\xi), V_1(\xi), U_2(\xi), V_2(\xi))\rangle = q(\xi, 1, 0, 0, 1), \tag{11.34}$$

in view of the initial conditions (8.9). We remark that an equation equivalent to that in (11.32) has been derived by Papanicolaou and Keller [**16**], by means of a two-variable expansion method, involving the fast variable x and the slow variable $\xi = \varepsilon^2 x$.

It is a straightforward matter to calculate the first- and second-order moments of U_1, V_1, U_2 and V_2 from (11.32)–(11.34), and this was done in [**18**]. It was found that the results agree with the perturbation results obtained in §§8 and 9. It is remarked that the limit theorem is not strictly applicable to the calculation of the moments, since it leads to unbounded functions $G(Y, \theta)$. The correlation functions were also calculated in [**18**], by introducing the conditional moment

$$g(\xi, Y, x; \mathbf{Z}, z) = \int G(\boldsymbol{\eta}, x + \xi/\varepsilon^2; \mathbf{Z}, z) p(\xi, \boldsymbol{\eta}; \mathbf{Y})\, d\boldsymbol{\eta}, \tag{11.35}$$

corresponding to (11.28). Then g satisfies equation (11.29), and

$$g(0, \mathbf{Y}, x; \mathbf{Z}, z) = G(\mathbf{Y}, x; \mathbf{Z}, z). \tag{11.36}$$

It follows from (11.35) that

$$\begin{aligned} &\langle g(\xi, \mathbf{Y}(\zeta), \zeta/\varepsilon^2; \mathbf{Y}(\zeta), \zeta/\varepsilon^2)\rangle \\ &\qquad = \langle G(\mathbf{Y}(\xi + \zeta), (\xi + \zeta)/\varepsilon^2; \mathbf{Y}(\zeta), \zeta/\varepsilon^2)\rangle. \end{aligned} \tag{11.37}$$

By this means, the approximate relationship (9.31), expressing the correlation functions in terms of the first- and second-order moments, was verified in [**18**].

In considering expectations of other functions of U_1, V_1, U_2 and V_2, a transformation of variables was made in [**18**], which considerably simplified the problem. Thus, let

$$\left(U_1 - \frac{i}{\beta_0} V_1\right) = \left(\frac{\sigma}{2}\right)^{1/2} [(z + 1)^{1/2} e^{i\alpha} + (z - 1)^{1/2} e^{-i\theta}], \tag{11.38}$$

$$\left(U_2 - \frac{i}{\beta_0} V_2\right) = \frac{iD}{\beta_0 (2\sigma)^{1/2}} [(z - 1)^{1/2} e^{-i\theta} - (z + 1)^{1/2} e^{i\alpha}], \tag{11.39}$$

where $z \geqq 1$, α and θ are angles, D is a real variable, and $\sigma > 0$ is a parameter to be chosen suitably. The positive square roots are to be taken in (11.38) and (11.39). Then, after some laborious calculations, it was found, from (11.32), with $q = f$, that

$$\begin{aligned} \frac{\partial f}{\partial \xi} &= (\kappa + \beta_0/\varepsilon^2)\left(\frac{\partial f}{\partial \alpha} - \frac{\partial f}{\partial \theta}\right) + \frac{1}{2}(c + a)\left(\frac{\partial}{\partial \alpha} - \frac{\partial}{\partial \theta}\right)^2 f \\ &\quad + 2a \frac{\partial}{\partial z}\left[(z^2 - 1)\frac{\partial f}{\partial z}\right] + \frac{a}{(z - 1)} \frac{\partial^2 f}{\partial \theta^2} - \frac{a}{(z + 1)} \frac{\partial^2 f}{\partial \alpha^2}. \end{aligned} \tag{11.40}$$

Corresponding to (11.33) the initial condition is

$$f(0, z, D, \alpha, \theta) = F(z, D, \alpha, \theta). \tag{11.41}$$

Note that no derivatives with respect to D occur in (11.40). This is not surprising since, from (11.38) and (11.39),

$$D = (U_1V_2 - V_1U_2), \tag{11.42}$$

and so D corresponds to the Wronskian of the solutions, and hence is constant. From the initial conditions (8.9) it follows that we are interested only in $D = 1$, and corresponding to (11.34) we have, from (11.38) and (11.39),

$$\langle F(z(\xi), 1, \alpha(\xi), \theta(\xi))\rangle = f(\xi, \tfrac{1}{2}(\sigma + 1/\sigma), 1, 0, \pi H(\sigma - 1)), \tag{11.43}$$

where H denotes Heaviside's step function. It is remarked that if $\sigma = 1$ then the initial value of θ is arbitrary.

The mean power transmitted by a plane electromagnetic wave normally incident on a randomly stratified dielectric slab was calculated in [**18**], with the help of the above formulation. It was assumed that the slab fills the region $0 \leqq x \leqq L$, in which the dielectric constant is $K(x) = n_0^2[1 + \varepsilon N(x)]$, where $N(x)$ is a bounded stochastic process satisfying (8.10). Thus n_0^2 is the average dielectric constant of the slab, and $\beta_0 = k_0 n_0$, in (8.8), where k_0 is the free space wave number. The regions $x < 0$ and $x > L$ are filled with nonstochastic dielectric media having the dielectric constants $\mu^2 n_0^2$ and $v^2 n_0^2$, respectively, where $\mu > 0$ and $v > 0$. If the wave is incident at $x = 0$, then the amplitude transmission coefficient T satisfies [**8**]

$$|T|^2 = 4\mu^2[(2\mu v + v_1^2\beta_0^{-2} + v^2u_1^2 + \mu^2v_2^2 + \mu^2v^2\beta_0^2u_2^2)^{-1}]_{x=L}. \tag{11.44}$$

If we take $D = 1$ and $\sigma = \mu$ in (11.38) and (11.39), then the initial condition in (11.41) corresponding to $|T|^2$ is

$$\begin{aligned} &f(0, z, 1, \alpha, \theta) \\ &\quad = 4\mu[2v + (1 + v^2)z - (1 - v^2)(z^2 - 1)^{1/2}\cos(\alpha - \theta)]^{-1}, \end{aligned} \tag{11.45}$$

and, from (11.31) and (11.43),

$$\langle |T|^2\rangle \approx f(\varepsilon^2 L, \tfrac{1}{2}(\mu + 1/\mu), 1, 0, \pi H(\mu - 1)). \tag{11.46}$$

The solution of (11.40), satisfying the initial condition (11.45), was obtained by expanding in a Fourier series in $(\alpha - \theta)$, and using Mehler transforms [**49**]. The reader is referred to [**18**] for the details.

We finally consider the general case, wherein the function F in (11.41) is such that the limit theorem is applicable (which is the case for $|T|^2$), and we determine the slowly varying part of the mean in (11.43). We now take $D = 1$ and $\sigma = 1$ in (11.38) and (11.39), and expand f in a double Fourier series,

$$f(\xi, z, 1, \alpha, \theta) = \sum_{m=-\infty}^{\infty} \sum_{n=-\infty}^{\infty} f_{mn}(\xi, z)\, e^{i(m\alpha + n\theta)}. \tag{11.47}$$

Then, from (11.40),

$$\begin{aligned}\frac{\partial f_{mn}}{\partial \xi} &= 2a \frac{\partial}{\partial z}\left[(z^2 - 1)\frac{\partial f_{mn}}{\partial z}\right] + a\left[\frac{m^2}{(z+1)} - \frac{n^2}{(z-1)}\right] f_{mn} \\ &\quad - \tfrac{1}{2}(c + a)(m - n)^2 f_{mn} + i(\kappa + \beta_0/\varepsilon^2)(m - n) f_{mn},\end{aligned} \tag{11.48}$$

and, from (11.41), the initial condition is

$$f_{mn}(0, z) = F_{mn}(z) \equiv \frac{1}{4\pi^2}\int_0^{2\pi}\int_0^{2\pi} F(z, 1, \alpha, \theta)\, e^{-i(m\alpha + n\theta)}\, d\alpha\, d\theta. \tag{11.49}$$

If $f_{mn}(\xi, z)$ is a bounded solution of equation (11.48), then it follows that $f_{mn}(\xi, 1) = 0$ unless $n = 0$. Hence, from (11.43) and (11.47), with $\sigma = 1$,

$$\langle F(z(\xi), 1, \alpha(\xi), \theta(\xi))\rangle = \sum_{m=-\infty}^{\infty} f_{m0}(\xi, 1). \tag{11.50}$$

It also follows from (11.48) that f_{mn} is a rapidly varying function of ξ unless $m = n$. Thus, from (11.50), the slowly varying part of $\langle F\rangle$ is $\langle F\rangle_0 = f_{00}(\xi, 1)$. The equation for f_{00} may be solved by means of Mehler transforms [**49**], and it was found in [**18**], for suitable $F_{00}(z)$, that

$$\begin{aligned}\langle F\rangle_0 = e^{-a\xi/2}\int_0^{\infty} &\exp(-2a\xi y^2) y \\ &\cdot \tanh(\pi y)\int_1^{\infty} F_{00}(z)\mathscr{P}_{-1/2+iy}(z)\, dz\, dy,\end{aligned} \tag{11.51}$$

where $\mathscr{P}_{-1/2+iy}(z)$ is a conical Legendre function [**50**]. An alternate representation is [**21**]

$$\langle F\rangle_0 = \frac{e^{-a\xi/2}}{2\pi^{1/2}(a\xi)^{3/2}}\int_0^{\infty} \rho \exp\left(\frac{-\rho^2}{2a\xi}\right)\int_1^{\cosh 2\rho} \frac{F_{00}(z)\, dz}{(\cosh 2\rho - z)^{1/2}}\, d\rho. \tag{11.52}$$

12. **Perturbation results for coupled lines.** In this concluding section we apply some results of Papanicolaou and Keller [**16**] to some coupled line equations. They investigated perturbed stochastic differential equations, by means of a two-variable procedure, and we begin by stating their results in a form which will suffice for our purposes. Thus, consider the linear vector stochastic differential equation

$$dw/dx = [\varepsilon A_1(x) + \varepsilon^2 A_2] w(x), \qquad w(0) = w_0, \tag{12.1}$$

where $A_1(x)$ is a random matrix with zero mean,

(12.2) $$\langle A_1(x)\rangle \equiv 0,$$

A_2 is a constant nonstochastic matrix, and w_0 is a nonstochastic vector. Define

(12.3) $$(A_1A_1)^- = \lim_{X\to\infty}\left[\frac{1}{X}\int_0^X\int_0^x \langle A_1(x)A_1(\xi)\rangle\, d\xi\, dx\right].$$

Then [16] asymptotically, for $0 < \varepsilon \ll 1$,

(12.4) $$d\langle w(x)\rangle/dx \approx \varepsilon^2((A_1A_1)^- + A_2)\langle w(x)\rangle, \qquad \langle w(0)\rangle = w_0.$$

We now consider the coupled line equations (3.26), subject to the initial conditions (3.27). The quantities of interest are $R_{kl}(x)$, as given by (3.28). We suppose that $N(x)$ is a real, wide sense stationary stochastic process satisfying

(12.5) $$\langle N(x)\rangle = 0, \qquad \langle N(x)N(\xi)\rangle = \rho(x-\xi).$$

For sufficiently smooth $N(x)$, the quantities $r_{kl}(x)$ defined in (3.33) satisfy the equations

(12.6) $$dr_{00}/dx = iN(x)(C_\sigma r_{10} - C_0 r_{01}),$$

(12.7) $$dr_{01}/dx = iN(x)(C_\sigma r_{11} - C_0 r_{00}) + \Delta\gamma^* r_{01},$$

(12.8) $$dr_{10}/dx = iN(x)(C_\sigma r_{00} - C_0 r_{11}) + (\Delta\gamma + i\sigma)r_{10},$$

(12.9) $$dr_{11}/dx = iN(x)(C_\sigma r_{01} - C_0 r_{10}) + (\Delta\gamma^* + \Delta\gamma + i\sigma)r_{11},$$

where C_0 and C_σ are given by (3.34). The initial conditions are

(12.10) $$r_{kl}(0) = i_k i_l^* \qquad (k, l = 0, 1).$$

Note, from (3.28) and (3.33), that

(12.11) $$R_{kl}(x) = \langle r_{kl}(x)\rangle.$$

In [3] we investigated the system (12.6)–(12.10), both in the nonresonance and resonance cases. Thus, we assumed that there is a characteristic length l, e.g., the correlation length, and an ε, with $0 < \varepsilon \ll 1$, such that

(12.12) $$C_0 l = O(\varepsilon), \qquad C_\sigma l = O(\varepsilon), \qquad \Delta\alpha l = O(\varepsilon^2), \qquad \sigma l = O(\varepsilon^2),$$

where $\Delta\alpha$ is given by (3.23). In the nonresonance case

(12.13) $$\Delta\beta l = O(1).$$

This can best be described as a case with weak coupling, weak attenuation and narrow fractional bandwidth. We consider the asymptotic calculation of $R_{kl}(x)$, valid for $0 \leqq x/l \leqq O(1/\varepsilon^2)$. Thus, we let

$$w = (r_{00}, r_{11}, e^{i\Delta\beta x} r_{01}, e^{-i\Delta\beta x} r_{10})^t, \tag{12.14}$$

where t denotes transpose. Then, from (3.23), (12.12) and (12.13), the system (12.6)–(12.10) takes the form (12.1), with

$$\varepsilon A_1(x) = iN(x)\begin{bmatrix} 0 & 0 & -C_0 e^{-i\Delta\beta x} & C_\sigma e^{i\Delta\beta x} \\ 0 & 0 & C_\sigma e^{-i\Delta\beta x} & -C_0 e^{i\Delta\beta x} \\ -C_0 e^{i\Delta\beta x} & C_\sigma e^{i\Delta\beta x} & 0 & 0 \\ C_\sigma e^{-i\Delta\beta x} & -C_0 e^{-i\Delta\beta x} & 0 & 0 \end{bmatrix}, \tag{12.15}$$

and

$$\varepsilon^2 A_2 = \begin{bmatrix} 0 & 0 & 0 & 0 \\ 0 & (2\Delta\alpha + i\sigma) & 0 & 0 \\ 0 & 0 & \Delta\alpha & 0 \\ 0 & 0 & 0 & (\Delta\alpha + i\sigma) \end{bmatrix}. \tag{12.16}$$

Note that $A_1(x)$ as given by (12.15) satisfies (12.2), by virtue of (12.5). We define

$$S(\omega) = \int_0^\infty e^{i\omega\xi} \rho(\xi)\, d\xi, \tag{12.17}$$

and assume that

$$\lim_{X\to\infty}\left[\frac{1}{X}\int_0^X \xi|\rho(\xi)|\, d\xi\right] = 0. \tag{12.18}$$

Then, for real ω,

$$\begin{aligned} &\lim_{X\to\infty}\left[\frac{1}{X}\int_0^X \int_0^x e^{i\omega(x-\xi)} \rho(x-\xi)\, d\xi\, dx\right] \\ &\qquad = \lim_{X\to\infty}\left[\frac{1}{X}\int_0^X (X-\xi)e^{i\omega\xi} \rho(\xi)\, d\xi\right] = S(\omega). \end{aligned} \tag{12.19}$$

Also,

$$\int_0^X \int_0^x e^{i\omega(x+\xi)} \rho(x-\xi)\, d\xi\, dx = \frac{1}{2i\omega}\int_0^X [e^{i\omega(2X-\xi)} - e^{i\omega\xi}]\rho(\xi)\, d\xi. \tag{12.20}$$

Hence, for real $\omega \neq 0$,

$$\lim_{X\to\infty}\left[\frac{1}{X}\int_0^X\int_0^x e^{i\omega(x+\xi)}\,\rho(x-\xi)\,d\xi\,dx\right]=0. \tag{12.21}$$

From (12.3)–(12.5), (12.11), (12.15), (12.16), (12.19) and (12.21), it is found that

$$\begin{aligned} dR_{00}/dx \approx {} & -[C_0^2S(-\Delta\beta)+C_\sigma^2S(\Delta\beta)]R_{00} \\ & +C_0C_\sigma[S(\Delta\beta)+S(-\Delta\beta)]R_{11}, \end{aligned} \tag{12.22}$$

$$\begin{aligned} dR_{11}/dx \approx {} & C_0C_\sigma[S(-\Delta\beta)+S(\Delta\beta)]R_{00} \\ & +[(2\Delta\alpha+i\sigma)-C_0^2S(\Delta\beta)-C_\sigma^2S(-\Delta\beta)]R_{11}, \end{aligned} \tag{12.23}$$

and

$$dR_{01}/dx \approx [(\Delta\alpha - i\Delta\beta)-(C_0^2+C_\sigma^2)S(\Delta\beta)]R_{01}, \tag{12.24}$$

$$dR_{10}/dx \approx [(\Delta\alpha + i\Delta\beta + i\sigma)-(C_0^2+C_\sigma^2)S(-\Delta\beta)]R_{10}. \tag{12.25}$$

From (12.10) and (12.11), the initial conditions are

$$R_{kl}(0)=i_ki_l^*. \tag{12.26}$$

For $\rho(\xi)=D_0\delta(\xi)$ we have, from (12.17), $S(\omega)\equiv\frac{1}{2}D_0$. Thus, as is seen from (3.23), (3.39) and (3.40), the asymptotic equations for R_{00} and R_{11} are exact for white noise. However, from (12.24)–(12.26), for $S(\omega)\equiv\frac{1}{2}D_0$, we have asymptotically

$$R_{01}(x)\approx i_0i_1^*\exp\{[\Delta\gamma^*-\tfrac{1}{2}(C_0^2+C_\sigma^2)D_0]x\}, \tag{12.27}$$

$$R_{10}(x)\approx i_1i_0^*\exp\{[(\Delta\gamma+i\sigma)-\tfrac{1}{2}(C_0^2+C_\sigma^2)]x\}. \tag{12.28}$$

It is readily verified that these results are asymptotically consistent for $0\leqq x/l\leqq O(1/\varepsilon^2)$, to the lowest order in ε, with the exact equations (3.41) and (3.42) for white noise. The characteristic roots corresponding to those equations are

$$\begin{aligned} \lambda &= (\Delta\alpha+\tfrac{1}{2}i\sigma)-\tfrac{1}{2}(C_0^2+C_\sigma^2)D_0 \pm [C_0^2C_\sigma^2D_0^2-(\Delta\beta+\tfrac{1}{2}\sigma)^2]^{1/2} \\ &= \pm i(\Delta\beta+\tfrac{1}{2}\sigma)+(\Delta\alpha+\tfrac{1}{2}i\sigma)-\tfrac{1}{2}(C_0^2+C_\sigma^2)D_0+O(\varepsilon^4)/l, \end{aligned} \tag{12.29}$$

from (12.12) and (12.13).

The above results are valid under the assumptions (12.12) and (12.13). Suppose now that instead of (12.13) we have

$$\Delta\beta l = O(\varepsilon^2), \tag{12.30}$$

which we refer to as the resonance case. Note that (12.12) and (12.30) may also correspond to the case of large coupling, moderate attenuation and fractional bandwidth, and short correlation length, with $l = \varepsilon^2 l_0$. We now take

$$w = (r_{00}, r_{11}, r_{01}, r_{10})^t. \tag{12.31}$$

Then the system (12.6)–(12.11) has the form (12.1), with

$$\varepsilon A_1(x) = iN(x) \begin{bmatrix} 0 & 0 & -C_0 & C_\sigma \\ 0 & 0 & C_\sigma & -C_0 \\ -C_0 & C_\sigma & 0 & 0 \\ C_\sigma & -C_0 & 0 & 0 \end{bmatrix}, \tag{12.32}$$

and

$$\varepsilon^2 A_2 = \begin{bmatrix} 0 & 0 & 0 & 0 \\ 0 & (\Delta\gamma^* + \Delta\gamma + i\sigma) & 0 & 0 \\ 0 & 0 & \Delta\gamma^* & 0 \\ 0 & 0 & 0 & (\Delta\gamma + i\sigma) \end{bmatrix}. \tag{12.33}$$

Note that $A_1(x)$ as given by (12.32) satisfies (12.2), by virtue of (12.5). From (12.3)–(12.5), (12.11), (12.19) and (12.31)–(12.33), it is found that

$$dR_{00}/dx \approx -(C_0^2 + C_\sigma^2)S(0)R_{00} + 2C_0C_\sigma S(0)R_{11}, \tag{12.34}$$

$$\begin{aligned} dR_{11}/dx \approx{}& 2C_0C_\sigma S(0)R_{00} \\ &+ [(\Delta\gamma^* + \Delta\gamma + i\sigma) - (C_0^2 + C_\sigma^2)S(0)]R_{11}, \end{aligned} \tag{12.35}$$

and

$$dR_{01}/dx \approx [\Delta\gamma^* - (C_0^2 + C_\sigma^2)S(0)]R_{01} + 2C_0C_\sigma S(0)R_{10}, \tag{12.36}$$

$$dR_{10}/dx \approx 2C_0C_\sigma S(0)R_{01} + [(\Delta\gamma + i\sigma) - (C_0^2 + C_\sigma^2)S(0)]R_{10}. \tag{12.37}$$

We remark that the difference in structure between equations (12.36) and (12.37), and equations (12.24) and (12.25), arises because of the discontinuity of the limit in (12.21) at $\omega = 0$. In the resonance case, the asymptotic equations (12.34)–(12.37) are exact for white noise, with $S(0) = \frac{1}{2}D_0$, as is seen from (3.39)–(3.42). We comment that white noise has zero correlation length.

In conclusion, we remark that D. Marcuse [**51**] has obtained equations for the average powers in n coupled lines, with weak random coupling. He gave a somewhat formal derivation, but the resulting equations may be obtained by the procedure discussed in this section. The details have been given by G. C. Papanicolaou [**52**].

Acknowledgements. For many helpful comments and discussions we are grateful to our colleagues V. E. Benes, T. T. Kadota, S. P. Lloyd, D. Slepian, H. S. Witsenhausen and especially L. A. Shepp. We have also benefited from discussions with G. C. Papanicolaou and J. B. Keller. Lastly, we wish to thank Mrs. J. King and Miss J. Langoski for their efficient typing of this manuscript.

REFERENCES

1. J. L. Doob, *Stochastic processes*, Wiley, New York; Chapman & Hall, London, 1953. MR **15**, 445.

2. H. P. McKean, *Stochastic integrals*, Probability and Math. Statist., no. 5, Academic Press, New York, 1969, p. 33. MR **40** #947.

3. J. A. Morrison and J. McKenna, *Coupled line equations with random coupling*, Bell System Tech. J. **51** (1972), 209–228.

4. H. E. Rowe, *Propagation in one-dimensional random media*, IEEE Trans. on Microwave Theory and Techniques **19** (1971), 73–80.

5. H. E. Rowe and D. T. Young, *Transmission distortion in multi-mode random waveguides*, IEEE Trans. on Microwave Theory and Techniques **20** (1972), 349–365.

6. D. T. Young and H. E. Rowe, *Optimum coupling for random guides with frequency-dependent coupling*, IEEE Trans. on Microwave Theory and Techniques **20** (1972), 365–372.

7. J. E. Molyneux, *Wave propagation in certain one-dimensional random media*, J. Mathematical Phys. **13** (1972), 58–69.

8. J. McKenna and J. A. Morrison, *Moments and correlation functions of solutions of a stochastic differential equation*, J. Mathematical Phys. **11** (1970), 2348–2360. MR **42** #1232.

9. ———, *Moments of solutions of a class of stochastic differential equations*, J. Mathematical Phys. **12** (1971), 2126–2136.

10. J. B. Keller, *Wave propagation in random media*, Proc. Sympos. Appl. Math., vol. 13, Amer. Math. Soc., Providence, R.I., 1962, pp. 227–246. MR **25** #3683.

11. ———, *Stochastic equations and wave propagation in random media*, Proc. Sympos. Appl. Math., vol. 16, Amer. Math. Soc., Providence, R.I., 1964, pp. 145–170. MR **31** #2895.

12. R. C. Bourret, *Stochastically perturbed fields, with application to wave propagation in random media*, Nuovo Cimento (10) **26** (1962), 1–31. MR **26** #2276.

13. ———, *Propagation of randomly perturbed fields*, Canad. J. Phys. **40** (1962), 782–790. MR **27** #6473.

14. ———, *Ficton theory of dynamical systems with noisy parameters*, Canad. J. Phys. **43** (1965), 619–639. MR **31** #1062.

15. ———, *Stochastic systems equivalent to second quantized systems; examples*, Canad. J. Phys. **44** (1966), 2519–2524.

16. G. C. Papanicolaou and J. B. Keller, *Stochastic differential equations with applications to random harmonic oscillators and wave propagation in random media*, SIAM J. Appl. Math. **21** (1971), 287–305.

17. R. Z. Has'minskiĭ, *A limit theorem for solutions of differential equations with a random right hand part*, Teor. Verojatnost. i Primenen. **11** (1966), 444–462 = Theor. Probability Appl. **11** (1966), 390–406. MR **34** #3637.

18. J. A. Morrison, *Application of a limit theorem to solutions of a stochastic differential equation*, J. Math. Anal. Appl. **39** (1972), 13–36.

19. M. Lax, *Classical noise*. IV: *Langevin methods*, Rev. Modern Phys. **38** (1966), 541–566.

20. G. C. Papanicolaou, *Wave propagation in a one-dimensional random medium*, SIAM J. Appl. Math. **21** (1971), 13–18.

21. J. A. Morrison, G. C. Papanicolaou and J. B. Keller, *Mean power transmission through a slab of random medium*, Comm. Pure Appl. Math. **24** (1971), 473–489.

22. U. Frisch, *Wave propagation in random media*, Probabilistic Methods in Appl. Math., vol. 1, Academic Press, New York, 1968, pp. 75–198. MR **42** #4088.

23. M. Fibich and E. Helfand, *Statistical properties of waves in a random medium*, Phys. Rev. (2) **183** (1969), 265–277. MR **40** #8409.

24. W. B. Davenport, Jr. and W. L. Root, *An introduction to the theory of random signals and noise*, McGraw-Hill, New York, 1958. MR **19,** 1090.

25. J. A. McFadden, *The probability density of the output of a filter when the input is a random telegraphic signal*, IRE National Convention Record **7** (1959), 164–169.

26. D. A. Darling, *On a class of problems related to the random division of an interval*, Ann. Math. Statist. **24** (1953), 239–253. MR **15,** 444.

27. J. McKenna and J. A. Morrison, *Exact solutions to some deterministic and random transmission line problems*, Bell System Tech. J. **51** (1972), 1269–1292.

28. U. Grenander, *Some non linear problems in probability theory*, Probability and Statistics: The Harald Cramér Volume, Almqvist & Wiksell, Stockholm; Wiley, New York, 1959. MR **22** #247.

29. R. E. Bellman, *Introduction to matrix analysis*, McGraw-Hill, New York, 1960. MR **23** #A153.

30. W. M. Wonham, *Optimal stationary control of a linear system with state-dependent noise*, SIAM J. Control **5** (1967), 486–500. MR **36** #2421.

31. S. E. Miller, *Coupled wave theory and waveguide applications*, Bell System Tech. J. **33** (1954), 661–720.

32. S. A. Schelkunoff, *Conversion of Maxwell's equations into generalized telegraphist's equations*, Bell System Tech. J. **34** (1955), 995–1043. MR **18,** 969.

33. W. H. Louisell, *Coupled mode and parametric electronics*, Wiley, New York, 1960.

34. A. L. Jones, *Coupling of optical fibers and scattering in fibers*, J. Opt. Soc. Amer. **55** (1965), 261–271.

35. D. Marcuse, *Mode conversion caused by surface imperfections of a dielectric slab waveguide*, Bell System Tech. J. **48** (1969), 3187–3215.

36. S. E. Miller, *Waveguide as a communication medium*, Bell System Tech. J. **33** (1954), 1209–1265.

37. H. E. Rowe and W. A. Warters, *Transmission in multimode waveguide with random imperfections*, Bell System Tech. J. **41** (1962), 1031–1170.

38. W. M. Wonham, *Random differential equations in control theory*, Probabilistic Methods in Appl. Math., vol. 2, Academic Press, New York, 1970, pp. 131–212. MR **41** #6345.

39. N. N. Krasovskiĭ and E. A. Lidskiĭ, *Analytic design of controllers in systems with random attributes*. I, II, III, Avtomat. i Telemeh. **22** (1961), 1145–1150, 1273–1278, 1425–1431 = Automat. Remote Control **22** (1962), 1021–1025, 1141–1146, 1289–1294. MR **27** #2375; #2376; #2377.

40. ———, *Analytic design of controllers in stochastic systems with velocity-limited controlling action*, Prikl. Mat. Meh. **25** (1961), 420–432 = J. Appl. Math. Mech. **25** (1961), 627–643. MR **25** #1086.

41. J. A. Morrison, *Moments and correlation functions of solutions of some stochastic matrix differential equations*, J. Mathematical Phys. **13** (1972), 299–306.

42. ———, *Calculation of correlation functions of solutions of a stochastic ordinary differential equation*, J. Mathematical Phys. **11** (1970), 3200–3209. MR **42** #6954.

43. C. J. Burke and M. Rosenblatt, *A Markovian function of a Markov chain*, Ann. Math. Statist. **29** (1958), 1112–1122. MR **21** #367.

44. J. A. Morrison and J. McKenna, *Application of the smoothing method to a stochastic ordinary differential equation*, J. Mathematical Phys. **11** (1970), 2361–2367. MR **42** #1234.

45. W. Feller, *An introduction to probability theory and its application*. Vol. 2, Wiley, New York, 1966, p. 586. MR **35** #1048.

46. J. B. Keller, *A survey of the theory of wave propagation in continuous random media*, Proc. Sympos. on Turbulence of Fluids and Plasmas, Brooklyn Poly. Inst., Brooklyn, N.Y., 1968, pp. 131–142.

47. J. J. McCoy, *Higher order moments of the inverse of a linear stochastic operator*, J. Opt. Soc. Amer. **62** (1972), 30–40.

48. W. Feller, p. 243 of [**45**].

49. W. Magnus, F. Oberhettinger and R. P. Soni, *Formulas and theorems for the special functions of mathematical physics*, 3rd ed., Die Grundlehren der math. Wissenschaften, Band 52, Springer-Verlag, New York, 1966, p. 398. MR **38** #1291.

50. ———, p. 201 of [**49**].

51. D. Marcuse, *Derivation of coupled power equations*, Bell System Tech. J. **51** (1972), 229–237.

52. G. C. Papanicolaou, *A Kinetic theory for power transfer in stochastic systems*, J. Mathematical Phys. **13** (1972), 1912–1918.

53. Since going to press, it has come to our attention that A. Brissaud and U. Frisch have given an alternate direct proof of this result (private communication).

Bell Laboratories, Murray Hill

Optimal Control of Diffusion Processes

Wendell H. Fleming[1]

1. **Introduction.** This paper summarizes some recent work on optimal control theory for continuous parameter stochastic processes. We discuss only the control of Markov diffusion processes ξ governed by stochastic differential equations of Itô type. Moreover, we consider only the two cases when either: (A) no observations are available to the controller (open-loop control); or (B) the states of the process ξ are completely observed by the controller. The optimal control of partially observable processes is discussed by Varaiya [**V**] elsewhere in this volume. A dynamic programming treatment of optimal control for jump Markov processes has recently been given by Rishel [**R2**].

For background about developments in optimal stochastic control theory up to about 1969, see for instance the survey [**F1**] and books by Astrom [**A**], Kushner [**K2**] and Wonham [**Wo**], and for some results about controlled 1-dimensional processes, Mandl [**Ma**].

Questions of interest for a theory of optimal control for diffusion processes include the following: existence of optimal controls; necessary and sufficient conditions for a minimum; qualitative features of the solution (continuity properties, special forms, etc.); relations with the theory of parabolic partial differential equations; and approximate computational methods. On the last question, it seems difficult to get results unless the state space is low-dimensional, the problem is linear-quadratic (or nearly so), or else the noise coefficient σ in the system equations (2.1) below is small.

The control problem which we shall formulate in §§2 and 3 is a pertur-

AMS (*MOS*) *subject classifications* (1970). Primary 93E20.

[1] This work was supported in part by the Air Force Office of Scientific Research under grant AF-AFOSR 71-2078 and in part by the National Science Foundation under grant GP-15132.

bation of the standard problem of Pontryagin type. The perturbation occurs by adding a noise term $\sigma\, dw$ in the system equations (2.1), and taking expected performance in (3.1).

As a fairly typical example, let us consider a harmonic oscillator driven by control plus white noise: $\ddot{\eta} + \eta = u + \sigma\dot{w}$, where $u(t)$ is the control applied at time t and w is a one-dimensional Brownian motion. Here $\dot{w} = d/dt$. The control could either be a function of t (the open-loop case), or a function of $(t, \eta, \dot{\eta})$. In the latter case we are using a control policy, or feedback control law, and are assuming that the displacement η and velocity $\dot{\eta}$ are observed at each t. Several kinds of criteria of system performance can be considered. If we suppose that control occurs on a fixed finite time interval $s \leqq t \leqq T$, then one might wish to minimize a positive quadratic criterion of the form

$$J = E\int_s^T (M\eta^2 + N\dot{\eta}^2 + u^2)\, dt.$$

For the case when the "state" $(\eta, \dot{\eta})$ is observed at each t, the method of dynamic programming applies. For the open-loop problem, other methods are mentioned in §4. Let us write $x_1 = \eta(s)$, $x_2 = \dot{\eta}(s)$ for the initial displacement and velocity, and $\phi(s, x_1, x_2)$ for the minimum expected cost for these initial data. Then ϕ is a solution of the partial differential equation (5.1), which in this example becomes

$$(*)\qquad \begin{aligned} &\frac{\partial\phi}{\partial s} + \frac{\sigma^2}{2}\frac{\partial^2\phi}{\partial x_2^2} + Mx_1^2 + Nx_2^2 \\ &\quad + x_2\frac{\partial\phi}{\partial x_1} - x_1\frac{\partial\phi}{\partial x_2} + \min_{u\in K}\left[u^2 + \frac{\partial\phi}{\partial x_2}u\right] = 0, \qquad s \leqq T, \end{aligned}$$

with the data $\phi(T, x) = 0$. Here K is an interval on the real line to which the control is constrained to lie. If there are no control constraints, $K = (-\infty, \infty)$, this is a special case of the linear regulator problem and the solution is well known. Then ϕ is a quadratic function of x_1, x_2; and the optimal control policy is a linear function of x_1, x_2 whose coefficients are computed from the solution of a matrix Ricatti equation. See for example [**K2**]. If we impose control constraints (say) $K = [-1, 1]$, then the problem is much more difficult and the explicit solution of (*) is not known. If we denote control policies by $Y(t, \eta, \dot{\eta})$ as in (2.2), then the half space $t \leqq T$ is divided into regions where $Y = -1$, $-1 < Y < 1$, $Y = 1$. Near the terminal hyperplane $t = T$, we must have $-1 < Y < 1$. A rather good approximation to the surface separating this region from a saturation region where $Y = \pm 1$ is known, when the noise coefficient σ is small [**F2**], [**H2**].

Another performance criterion often used is time to reach a preassigned target. In our harmonic oscillator example, we might take as a target the circle $\eta^2 + \dot{\eta}^2 = a^2$, and start with $x_1^2 + x_2^2 > a^2$. Let $\phi(x_1, x_2) =$ minimum expected time to reach the target. Then

$$(**) \qquad \frac{\sigma^2}{2}\frac{\partial^2\phi}{\partial x_1^2} + x_2\frac{\partial\phi}{\partial x_1} - x_1\frac{\partial\phi}{\partial x_2} + \min_{u\in K}\left[\frac{\partial\phi}{\partial x_2}u\right] = -1, \qquad x_1^2 + x_2^2 > a^2,$$

with $\phi = 0$ when $x_1^2 + x_2^2 = a^2$. For $\sigma = 0$ this is the Bushaw problem; in that case an exact solution is known [**HL**, p. 80]. The optimal control policy Y switches between $Y = 1$ and $Y = -1$. For small $\sigma > 0$ there is good approximation to the minimum expected time $\phi(x_1, x_2)$ to reach the target, but the precise way in which the optimal switching curve for the Bushaw problem perturbs for small $\sigma > 0$ is unknown. See [**H2**].

An interesting one-dimensional example arises in Merton's model of optimal portfolio selection [**Me**, §6]. Assume that there are two assets, one "risk free" with return rate r, and the other "risky" with price $P(t)$ fluctuating according to the stochastic differential equation $P^{-1}\,dP = \alpha\,dt + \sigma\,dw$. The control $u(t)$ consists of a pair $(\omega(t), C(t))$ where $\omega(t)$ is the fraction of wealth invested in the risky asset at time t and $C(t)$ the consumption rate at time t. The wealth $\xi(t)$ then fluctuates according to

$$d\xi = (1 - \omega)\xi r\,dt + \omega\xi(\alpha\,dt + \sigma\,dw) - C\,dt.$$

The problem is to maximize expected discounted total utility

$$J = E\int_s^T e^{-\rho t}V[C(t)]\,dt.$$

For $V(C)$ of the hyperbolic absolute risk aversion type, $V(C) = (aC + b)^\gamma$, it turns out that an optimal control policy gives consumption rate C and $\omega\xi$ as linear functions of ξ [**Me**, p. 391]. Another treatment of the portfolio selection problem, with a final-time cost criterion, is given in [**Wg**, §4].

2. **The system equations.** We consider the following model of a stochastic control process. The state process ξ evolves according to stochastic differential equations

$$(2.1) \qquad d\xi = f(t, \xi(t), u(t))\,dt + \sigma(t, \xi(t))\,dw, \qquad t \geqq s,$$

with initial data. Here w is a Brownian motion process. The state spaces of the processes ξ, w are finite-dimensional R^n, R^m for some n, m. We require that $u(t) \in K$, where the "control set" K is given. The process continues up to the time τ when $(t, \xi(t))$ exits from a given cylinder $Q = (T_0, T) \times B$, where $B \subset R^n$ is open and its boundary ∂B is a compact smooth manifold. Unless otherwise indicated, we assume that K is compact and convex, and that f, σ are smooth (class C^∞) and bounded together with their first-order partial derivatives. These assumptions can be weakened [**F1**, 5.5].

As already mentioned two cases will be considered.

(A) *Open-loop control.* In this case, the controls allowed are measurable functions from the time interval $[s, T]$ into the control set K. The initial data $\xi(s)$ is a given random vector independent of the w-process increments for times $\geqq s$, with $E|\xi(s)|^2 < \infty$.

(B) *States completely observed.* In particular, the initial state vector $x = \xi(s)$ is known. Now two kinds of controls can be admitted.

(i) *Control policies.* Such a policy is a function Y from Q into K, with suitable analytical properties insuring that (2.1) has a solution with the initial data, when we substitute

$$u(t) = Y(t, \xi(t)). \tag{2.2}$$

(ii) *Nonanticipative control processes.* Here u is a measurable stochastic process with state space K such that $u(t)$ is a $\mathscr{B}_t$ measurable random vector, where $\{\mathscr{B}_t\}$ is an increasing family of σ-algebras and $\mathscr{B}_t$ is independent of the w-process increments for times $\geqq t$.

The solution process ξ is well-defined, given any open-loop control $u(\cdot)$, or more generally any nonanticipative control process. We admit control policies Y which are Borel measurable. By K. Itô's criterion, the solution ξ of (2.1) is well-defined if, in addition, $Y(t, \cdot)$ satisfies a uniform Lipschitz condition. By using the Girsanov formula for transformation of measure, it can be shown that this Lipschitz condition on Y is unnecessary if the local covariance matrices $\sigma\sigma^T$ are uniformly positive definite (see [**SV**]). The same statement holds if the system splits as described in §5 below. Existence of the solution process ξ then follows as in [**R1**], and uniqueness of ξ in probability law as in [**NW**, Theorem 2].

3. **The criterion to be minimized.** Let

$$J = E\left\{ \int_s^\tau L(t, \xi(t), u(t))\, dt + \Phi(\tau, \xi(\tau)) \right\}. \tag{3.1}$$

Here L, Φ are given smooth functions, bounded together with their first-order partial derivatives. The problem is to find a control minimizing J. Clearly the open-loop minimum value is no less than the minimum using control policies, which in turn is no less than that obtained using nonanticipative control processes. It is intuitively reasonable, when the states $\xi(t)$ are completely observed, that the minimum among control policies is the same as among nonanticipative controls. This has been proved under the assumptions stated in §5.

4. **The open-loop control problem, fixed terminal time.** In this section the terminal time is fixed, $\tau \equiv T$. We mention several results.

(α) *Existence of an optimal open-loop control.* Besides the general assumptions above, assume

$$f(t, \xi, \cdot) \text{ is linear} \quad \text{and} \quad L(t, \xi, \cdot) \text{ is convex.} \tag{4.1}$$

Then existence of an optimal $u(\cdot)$ follows from [**FN**].

(β) *Necessary conditions for a minimum.* If $\sigma = \sigma(t)$, then a necessary condition rather similar to Pontryagin's principle was obtained by Kushner [**K1**]. A necessary condition which applies to a wider class of problems (including ones where $\sigma(t, \xi)$ is state-dependent) was later obtained in the thesis of V. Warfield [**W**]. She used the approach of McShane [**M**] to stochastic integration.

(γ) *Small noise coefficient results.* These appear in the thesis of Holland [**H1**]. We mention two kinds of results. Let $\sigma = (2\varepsilon)^{1/2}I$, where I is the identity matrix and $\varepsilon \geqq 0$ is small. In (3.1) we now write $J = J^\varepsilon(u; s, x)$ to indicate its dependence on ε, the control function $u = u(\cdot)$, and on the initial data $x = \xi(s)$.

Consider the minimum value

$$\psi^\varepsilon(s, x) = \min_{u(\cdot)} J^\varepsilon(u; s, x).$$

The first result asserts the differentiability of ψ^ε when $\varepsilon = 0$, under some assumptions. Suppose that $f(t, \xi, \cdot)$ is linear, and that the following strong convexity condition holds:

$$v^T L_{uu} v \geqq c|v|^2 \quad \text{for all } v, \tag{4.2}$$

where $c > 0$. Moreover assume that (s, x) is a regular point, in the sense that, for these initial data, the deterministic control problem ($\varepsilon = 0$) has a *unique* open-loop optimal control $u^0(\cdot)$. Then

$$\psi^\varepsilon(s, x) = \psi^0(s, x) + \varepsilon\chi(s, x) + o(\varepsilon), \tag{4.3}$$

where

$$\chi(s, x) = \int_s^T \Delta_x \psi^0(t, \xi^0(t))\, dt;$$

Δ_x denotes Laplacian in the "state variables" and ξ^0 is the solution of $d\xi^0 = f(t, \xi^0(t), u^0(t))\, dt$ with $\xi^0(s) = x$. The proof of (4.3) uses the fact that the expansion

$$J^\varepsilon = J^0 + \varepsilon(J^\varepsilon)'|_{\varepsilon=0} + o(\varepsilon)$$

holds uniformly with respect to $u(\cdot)$. If u^ε is optimal in the stochastic problem, with $\sigma = (2\varepsilon)^{1/2}I$, then the regularity of (s, x) is used together with the other assumptions to show that u^ε tends uniformly to u^0 as $\varepsilon \to 0$.

Other results of Holland concern the validity of expansions of the minimum value ψ^ε and of the optimal control u^ε in higher powers of ε:

$$(4.4)\qquad \psi^\varepsilon(s, x) = \psi^0(s, x) + \sum_{j=1}^{l} \varepsilon^j \chi_j(s, x) + o(\varepsilon^l),$$

$$(4.5)\qquad u^\varepsilon(t) = u^0(t) + \sum_{j=1}^{l} \varepsilon^j \mu_j(t) + o(\varepsilon^l)$$

with $\chi_1 = \chi$ in (4.3). In these results, it is assumed that the state variable ξ appears linearly in f, and that there are no control constraints ($K = R^d$ for some d). We mention here just one of these results. Suppose that

$$f(t, \xi, u) = A(t)\xi + B(t)u.$$

Using the separation principle, the open-loop stochastic control problem is then equivalent to a deterministic control problem with system equations

$$d\hat{\xi} = [A(t)\hat{\xi}(t) + B(t)u(t)]\,dt,$$

where $\hat{\xi}(t) = E\xi(t)$, $\hat{\xi}(s) = \xi(s) = x$, and where

$$J^\varepsilon = \int_s^T \hat{L}^\varepsilon(t, \hat{\xi}(t), u(t))\,dt$$

for suitable $\hat{L}^\varepsilon$. Moreover, assume that $L(t, \cdot, u)$ is a positive polynomial

$$L(t, \xi, u) = \sum_{|\alpha| \leqq m} g_\alpha(t, u)\xi^\alpha \geqq 0,$$

that $L(t, \cdot, \cdot)$ is convex on $R^n \times R^d$, and that (4.2) holds. Then the expansions (4.4)–(4.5) are valid for any $l = 1, 2, \ldots$.

5. **The problem with completely observed states.** In this case the method of dynamic programming is convenient. Let

$$\phi(s, x) = \min_Y J(s, x; Y).$$

If the matrices

$$a(s, x) = \tfrac{1}{2}\sigma(s, x)\sigma^T(s, x)$$

are uniformly positive definite, then ϕ satisfies the parabolic second-order partial differential equation of dynamic programming

$$(5.1)\qquad \frac{\partial \phi}{\partial s} + \sum_{i,j=1}^{n} a_{ij}(s, x)\frac{\partial^2 \phi}{\partial x_i \partial x_j} + \min_{u \in K}\left[L + \frac{\partial \phi}{\partial x} f\right] = 0$$

in Q, with boundary data

(5.2) $$\phi(s, x) = \Phi(s, x), \qquad (s, x) \in \partial Q - \{T_0\} \times B.$$

A policy Y is optimal if

(5.3) $$L(s, x, u) + \frac{\partial \phi}{\partial x} f(s, x, u) = \min \text{ on } K \quad \text{for } y = Y(s, x), \text{ almost everywhere in } Q.$$

In [**R1**], the uniform positive definiteness assumption about $a(s, x)$ is replaced by weaker assumptions which more often hold in applications. It is assumed in [**R1**] that the system equations (2.1) split:

(5.4) $$\begin{aligned} d\xi' &= f'(t, \xi(t))\, dt, \\ d\xi'' &= f''(t, \xi(t), u(t))\, dt + \sigma(t, \xi''(t))\, dw, \end{aligned}$$

where ξ' denotes the first n' components of ξ and ξ'' the last n'' components, $n' + n'' = n$. The matrices $\sigma\sigma^T$ are uniformly positive definite, of rank n''. Moreover, the solution of the equations obtained by replacing f'' by 0 in (5.4) is assumed to have a transition density, integrable to some power $\alpha > 1$ over $[s + h, T] \times B$ for any $h > 0$. The function ϕ is now a generalized solution of (5.1) in the sense that ϕ is Lipschitz and all partial derivatives of ϕ appearing in (5.1) are integrable to some power $p > 1$.

Small noise results. If $\sigma = (2\varepsilon)^{1/2} I$, let $\phi^\varepsilon(s, x)$ denote the minimum value of $J^\varepsilon(s, x; Y)$ and Y^ε the optimal control policy. Some results about expanding ϕ^ε and ϕ^ε_x in powers of ε up to first- or second-order terms appear in [**F2**]. These hold in regions where ϕ^0 is sufficiently smooth (note that ϕ^0 is a solution of the first-order equation obtained by setting $a_{ij} = 0$ in (5.1)). If there are no control constraints, then expansions of ϕ^ε and Y^ε in higher powers of ε (formally similar to (4.4), (4.5)) are obtained. In [**H2**] this method is applied to a stochastic linear regulator problem with saturation control constraints, and also to the stochastically perturbed Bushaw problem.

Other recent results. Optimal control of 1-dimensional diffusions with inaccessible boundaries was studied by Morton [**Mo**]. As a corollary of general methods for partially observable diffusions, a purely probabilistic treatment of the dynamic programming equation (5.1) was given in [**Da V**]. A nice approach to the problem of existence of an optimal nonanticipative control process has been taken in the papers [**B**], [**Du V**], [**NW**]. One hopes that this approach may ultimately yield existence theorems for the case of partial observations.

References

[**A**] K. J. Åström, *Introduction to stochastic control theory*, Math. in Sci. and Engineering, vol. 70, Academic Press, New York, 1970. MR **42** #5686.

[**B**] V. E. Benes, *Existence of optimal stochastic control laws*, SIAM J. Control **9** (1971), 446–472.

[**Da V**] M. H. A. Davis and P. Varaiya, *Dynamic programming conditions for partially observable systems* (preprint).

[**Du V**] T. Duncan and P. Varaiya, *On the solutions of a stochastic control system*, SIAM J. Control **9** (1971), 354–371.

[**F1**] W. H. Fleming, *Optimal continuous-parameter stochastic control*, SIAM Rev. **11** (1969), 470–509. MR **41** #9633.

[**F2**] ———, *Stochastic control for small noise intensities*, SIAM J. Control **9** (1971), 473–518.

[**FN**] W. H. Fleming and M. Nisio, *On the existence of optimal stochastic controls*, J. Math. Mech. **15** (1966), 777–794. MR **33** #7170.

[**H1**] C. Holland, *Small noise open loop control problems*, Ph.D. Thesis, Brown University, Providence, R.I., 1972.

[**H2**] ———, *A numerical technique for small noise stochastic control problems*, J. Optimization Theory Appl. (to appear).

[**HL**] H. Hermes and J. P. LaSalle, *Functional analysis and time optimal control*, Academic Press, New York, 1969.

[**K1**] H. J. Kushner, *On the stochastic maximum principle: Fixed time of control*, J. Math. Anal. Appl. **11** (1965), 78–92. MR **32** #3908.

[**K2**] ———, *Introduction to stochastic control theory*, Holt, Rinehart, Winston, New York, 1971.

[**M**] E. J. McShane, *Stochastic calculus and stochastic models* (to appear).

[**Ma**] P. Mandl, *Analytical treatment of one-dimensional Markov processes*, Die Grundlehren der math. Wissenschaften, Band 151, Springer-Verlag, New York; Academia, Prague, 1968. MR **40** #930.

[**Me**] R. C. Merton, *Optimum consumption and portfolio rules in a continuous-time model*, J. Economic Theory **3** (1971), 373–413.

[**Mo**] R. Morton, *On the optimal control of stationary diffusion processes with inaccessible boundaries and no discounting*, J. Appl. Probability **8** (1971), 551–560.

[**NW**] M. Nisio and S. Watanabe, *On the existence of optimal stochastic controls based on a complete observation* (preprint).

[**P**] M. L. Puterman, *On the optimal control of diffusion processes*, Stanford Univ., Dept. of Operations Research TR24, August 1972.

[**R1**] R. Rishel, *Weak solutions of a partial differential equation of dynamic programming*, SIAM J. Control **9** (1971), 519–528.

[**R2**] ———, *Optimality of controls for systems with jump Markov disturbances* (preprint).

[**SV**] D. W. Stroock and S. R. S. Varadhan, *Diffusion processes with continuous coefficients*. I, II, Comm. Pure Appl. Math. **22** (1969), 345–400, 479–530. MR **40** #6641; #8130.

[**V**] P. P. Varaiya, *Optimal control of partially observable processes*, SIAM-AMS Proc., vol. 6, Amer. Math. Soc., Providence, R.I., pp. 173–187.

[**W**] V. Warfield, *A stochastic maximum principle*, Thesis, Brown University, Providence, R.I., 1971.

[**Wg**] E. Wong, *Martingale theory and applications to stochastic problems in dynamical systems*, Res. Report 72/19, Dept. of Computing and Control, Imperial College of Sci. and Tech. London, June 1972.

[**Wo**] W. M. Wonham, *Random differential equations in control theory*, Probabilistic Methods in Appl. Math., vol. 2, Academic Press, New York, 1970. MR **41** #6345.

BROWN UNIVERSITY

Optimal Control of a Partially Observed Stochastic System

Pravin Varaiya[1]

Abstract. Necessary and sufficient conditions are presented for the optimal control of a stochastic process when only some of the components of the process are observed. The controlled process is modelled as a stochastic differential equation forced by Brownian motion. The solution of the differential equation is defined in such a way as to permit a very wide class of controls, and then Hamilton-Jacobi type of optimality criteria are derived.

1. **Introduction and summary.** This paper reviews the results presented in [**1**] concerning the optimal control of a system modelled by the stochastic differential equation

$$dz(t) = g(t, z(\cdot), u(t))\, dt + \sigma(t, z(\cdot))\, dB(t), \qquad 0 \leqq t \leqq 1,$$

$$z(0) = z_0,$$

where $z(\cdot)$ is the n-dimensional process to be controlled, $B(\cdot)$ is Brownian motion, and $u(\cdot)$ is the control. Let $y(t) \in R^m$ consist of the last m coordinates of $z(t)$. At each time t, the information available to the controller is the past history of the process $y(\cdot)$, i.e., $\{(\tau, y(\tau)) | 0 \leqq \tau \leqq t\}$, and based on this information, a control $u(t) \in U$ must be selected to minimize the expected cost

$$J(u) = E\left(\int_0^1 c(s, z(\cdot), u(s))\right) ds.$$

The first difficulty we face is to define a solution of the stochastic differential equation when controls of the form described above are permitted. This difficulty arises from the fact that it is not reasonable to

AMS (*MOS*) *subject classifications* (1970). Primary 49B25, 49C15, 93E20; Secondary 60G45, 60G35, 60H20.

[1] Research supported by the National Science Foundation under grant GK-10656X1 and by a fellowship from the Guggenheim Foundation.

restrict attention a priori to Lipschitz controls. It turns out however that by using the technique of transformation of measures due to Girsanov [**2**] we can define a solution for a wide class of controls. This is carried out in §2.

In §3 we turn our attention to the principle of optimality. The central notion here is that of the value function V_u of a control u which, for each time t, is the minimum expected future cost conditioned on the observations available at time t. This important notion is due to Rishel [**3**], and is related to earlier work of Dynkin [**4**]. The optimality principle essentially states that a control u^* is optimal if and only if its expected future cost conditioned on the available observations is equal to V_{u^*}. It then follows that V_{u^*} is a supermartingale.

§4 is devoted to the derivation of a representation for the supermartingale V_{u^*} in terms of the parameters of the process g, σ, and the observations. This representation is based on the important work of Meyer [**5**], and Kunita-Watanabe [**6**]. The representation permits us to rewrite the optimality principle in a form which resembles the well-known Hamilton-Jacobi equations which arise in the optimal control of completely observed diffusion processes [**7**].

The optimality conditions are specialized to the case of complete observations in §5. As a by-product we obtain the conditions for the diffusion case.

The proofs of various propositions in the paper are either omitted with a reference to the main technique involved or else an informal argument is given which lends itself easily (though perhaps laboriously) to mathematical rigor. The interested reader is referred to [**1**] for details. As a general comment we may say that the work in stochastic control has been limited until now to the case of diffusion processes with complete information, and the main approach has been via the theory of partial differential equations [**7**], [**8**]. Our approach is heavily probabilistic and permits us to consider a wider class of problems. The approach seems particularly useful in dealing with partial observations.

It may be helpful to the nonspecialist if some of the main concepts are illustrated in a simple problem. We take the special case of complete information covered in §5. The problem is to find a control function $u(t)$ depending upon the past of the trajectory of the one-dimensional differential equation

$$dz(t) = [az(t) + u(t)]\,dt + dB(t), \qquad 0 \leqq t \leqq 1,$$

$$z(0) = z_0,$$

so as to minimize the expected value of the cost

$$E\left(\int_0^1 bz^2(s) + u^2(s)\right) ds.$$

Let $u^*(t)$ be an optimal control and let $z(t)$ be the corresponding trajectory. Let $V(t, x)$ be the minimum expected value of the cost over the interval $[t, 1]$ when the system is started at time t in state x. By Theorem 6 there must exist two functions, denoted by $\wedge V(t, x)$ and $\nabla V(t, x)$, and a constant V^*, such that along the optimal trajectory $z(t)$ we have the representation

$$V(t, z(t)) = V^* + \int_0^t (\wedge V)(s, z(s))\, ds + \int_0^t (\nabla V)(s, z(s))\, dz(s).$$

Furthermore, the optimal control $u^*(t)$ must satisfy

$$\begin{aligned}(\wedge V)(t, z(t)) &+ \min_u \{(\nabla V)(t, z(t))(az(t) + u) + bz^2(t) + u^2\}\\ &= (\wedge V)(t, z(t)) + \{(\nabla V)(t, z(t))(az(t) + (u^*(t))^2) + bz^2(t) + (u^*(t))^2\}\\ &= 0.\end{aligned}$$

From this we get

$$u^*(t) = -\tfrac{1}{2}(\nabla V)(t, z(t))$$

and

$$(\wedge V)(t, z(t)) + \{(\nabla V)(t, z(t))(az(t)) + bz^2(t) - \tfrac{1}{4}((\nabla V)^2(t, z))^2\} = 0.$$

Next we suppose that $V(t, x)$ is sufficiently differentiable. Then from the remark following Theorem 6 we know that

$$(\wedge V)(t, x) = \frac{\partial V}{\partial t}(t, x) + \frac{1}{2}\frac{\partial^2 V}{\partial x^2}(t, x) \quad \text{and} \quad (\nabla V)(t, x) = \frac{\partial V}{\partial x}(t, x),$$

so the substitution in the last equation above leads us to the partial differential equation

$$\frac{\partial V}{\partial t} + \frac{1}{2}\frac{\partial^2 V}{\partial x^2} + a\frac{\partial V}{\partial x}x + bx^2 - \frac{1}{4}\left(\frac{\partial V}{\partial x}\right)^2 = 0.$$

From the definition of V we have the obvious boundary condition $V(1, x) \equiv 0$. If we try the trial solution $V(t, x) = \alpha(t)x^2 + \beta(t)$, and substitute in this partial differential equation, we obtain

$$\begin{aligned}\dot{\alpha}(t) + 2a\alpha(t) + b - \alpha^2(t) &= 0, & \dot{\beta}(t) + \alpha(t) &= 0,\\ \alpha(1) &= 0, & \beta(1) &= 0.\end{aligned}$$

Evidently we can solve this pair of ordinary differential equations and obtain the functions $\alpha(t)$, $\beta(t)$. The optional control u^* is then given by

$$u^*(t) = -\tfrac{1}{2}(\nabla V)(t, z(t)) = -\alpha(t)z(t),$$

and the minimum expected cost is

$$V(0, z_0) = \alpha(0)z_0^2 + \beta(0).$$

2. **Solution of the differential equation.** We wish to obtain the solution of the differential equation (d.e.)

$$dz(t) = g(t, z, u(t))\,dt + \sigma(t, z)\,dB(t), \qquad 0 \leqq t \leqq 1, \tag{1}$$
$$z(0) = z_0,$$

where $z_0 \in R^n$ is a fixed vector. We need to make certain assumptions on σ and g. To this end, let $w(t)$, $0 \leqq t \leqq 1$, be a standard n-dimensional Brownian motion process on a probability space $(\Omega, \mathscr{A}, \mu)$, and, for each t, let $\mathscr{A}_t$ be the sub-σ-algebra of $\mathscr{A}$ generated by the random variables $\{w(s) | 0 \leqq s \leqq t\}$. Also let C be the normed vector space of all continuous functions $z:[0, 1] \to R^n$ with norm $\|z\| = \max\{|z(t)| \,|0 \leqq t \leqq 1\}$. Let $\mathscr{F}$ be the Borel σ-algebra of C and let $\mathscr{F}_t$ be the sub-σ-algebra of $\mathscr{F}$ generated by the functions $\{z(s) | z \in C, 0 \leqq s \leqq t\}$.

We assume that σ enjoys the properties A_σ.

A_σ (i) The elements of σ, $\sigma_{ij}:[0, 1] \times C \to R$ are jointly measurable and $\{\sigma_{ij}(t)\}$ is adapted[2] to $\{\mathscr{F}_t\}$.

(ii) $\sigma(t, z)$ is nonsingular for all t, z.

(iii) There exists a solution $\{z(t)\}$ of the differential equation

$$dz(t) = \sigma(t, z)\,dw(t), \qquad 0 \leqq t \leqq 1, \tag{2}$$
$$z(0) = z_0,$$

such that $z(\cdot)$ has continuous sample paths, $\{z(t)\}$ is adapted to $\{\mathscr{A}_t\}$, and

$$\sum_{i,j} \int_0^t \sigma_{ij}^2(t, z)\,dt < \infty.$$

(iv) This solution is unique; i.e., if $\{y(t)\}$ is another solution, then both $\{z(t)\}$, $\{y(t)\}$ induce the same measure on $(C, \mathscr{F})$.[3]

Under A_σ the solution $\{z(t)\}$ of (2) defines a unique probability measure P on $(C, \mathscr{F})$, by $P(F) = \mu(z^{-1}(F))$. All the random variables we will

[2]Let $\theta(t): E \to R$, $0 \leqq t \leqq 1$, be a family of functions, and let $\mathscr{E}_t$ be a family of σ-algebras on E. $\{\theta(t)\}$ is said to be *adapted* to $\{\mathscr{E}_t\}$ if $\theta(t)$ is $\mathscr{E}_t$-measurable for each t.

[3] For conditions on σ which guarantee uniqueness see [**9**], [**10**].

encounter will be functions of this process $\{z(t)\}$, and hence we will take $(C, \mathscr{F}, P)$ as our basic probability space. Note that, since σ is nonsingular, $w(t)$ can also be regarded as a random variable on $(C, \mathscr{F}, P)$.

The control $u(t)$ is required to take values in a fixed constraint set U, a Borel subset of R^l. The function g is assumed to satisfy A_g.

A_g (i) $g:[0, 1] \times C \times R^l \to R^n$ is jointly measurable, and $\{g(t, \cdot, u)\}$ is adapted to $\{\mathscr{F}_t\}$ for each fixed $u \in U$.

(ii) There exists a function $g_0 : R \to R$ such that

$$|\sigma^{-1}(t, z)g(t, z, u)| \leqq g_0(\|z\|)$$

for all t, u. (It follows that $\int_0^1 |\sigma^{-1} g|^2 \, dt < \infty$.)

We are now ready to define the class of admissible controls. Let $y(t) \in R^m$ denote the last m components of $z(t)$ and let $\mathscr{Y}_t$ be the sub-σ-algebra of $\mathscr{F}_t$ generated by the functions $\{y(s) | z \in C, 0 \leqq s \leqq t\}$. Let $\mathscr{U}^0$ consist of all measurable functions $u:[0, 1] \times C \to U$ such that $\{u(t, \cdot)\}$ is adapted to $\{\mathscr{Y}_t\}$. For $u \in \mathscr{U}^0$, and $0 \leqq t \leqq 1$, let $g^u(t, z) = g(t, z, u(t, z))$ and define the positive random variable $\rho_s^t(u)$ by

$$\rho_s^t(u) = \exp\left(\int_0^t (\sigma^{-1} g^u)' \sigma^{-1} dz - \frac{1}{2} \int_0^t |\sigma^{-1} g^u|^2 \, d\tau\right),$$

where the first integral is to be interpreted as a stochastic integral. Let $\rho(u) = \rho_0^1(u)$. Note that $\rho_r^t(u) = \rho_r^s(u)\rho_s^t(u)$ for $r \leqq s \leqq t$.

THEOREM 1 (GIRSANOV [**2**]). *Let* $u \in \mathscr{U}^0$.

(i) $E(\rho(u)) = \int_C \rho(u)(z)P(dz) \leqq 1$.

(ii) *Suppose* $E(\rho(u)) = 1$. *Let* P^u *be the probability measure on* $(C, \mathscr{F})$ *defined by* $P^u(F) = \int_F \rho(u)P(dz)$. *Then the process* $B(t)$ *defined by*

$$(3) \qquad B(t) = \sigma^{-1}(t, z)\left[z(t) - z(0) - \int_0^t g^u(s)\, ds\right], \qquad 0 \leqq t \leqq 1,$$

is a Brownian motion on $(\Omega, \mathscr{F}, P^u)$.

(iii) *If* $E(\rho(u)) = 1$, *then* $E\{\rho_s^t(u) | \mathscr{F}_s\} = 1$ *a.s.* (P), $0 \leqq s \leqq t \leqq 1$.

Theorem 1 says that if $E(\rho(u)) = 1$, then the solution $\{z(t)\}$ of (2) on the probability space $(\Omega, \mathscr{F}, P)$ is also a solution of (1) on the probability space $(\Omega, \mathscr{F}, P^u)$. Let $\mathscr{U} = \{u \in \mathscr{U}^0 | E(\rho(u)) = 1\}$.[4] $\mathscr{U}$ will be our set of *admissible controls*. For $u \in \mathscr{U}$, we will take *the* solution of (1) to be the process $\{z(t)\}$ with probability law given by P^u. It can be shown that P^u is mutually absolutely continuous with respect to P. In particular then,

[4] The condition $E(\rho(u)) = 1$ is nontrivial. Various sufficient conditions are derived in [**11**], [**17**]. However no simple necessary and sufficient conditions are known.

if some property is stated to hold "almost surely" (a.s.), it is irrelevant which measure is being referred to. It can be shown further that the measure P^u induced by u on $(C, \mathscr{F})$ is unique. For details see [**11**].

Let $c:[0, 1] \times C \times R^l \to R$ be a nonnegative bounded measurable function such that $\{c(t, \cdot, u)\}$ is adapted to $\{\mathscr{F}_t\}$ for fixed u. The function c is called the cost function. If $u \in \mathscr{U}$, the risk or cost for u is

$$J(u) = E^u\left(\int_0^1 c^u(s)\,ds\right) = \int_C \left\{\int_0^1 c(s, z, u(s, z))\,ds\right\} P^u(dz).$$

Let $J^* = \inf\{J(u) | u \in \mathscr{U}\}$. The *optimal control problem* is to find $u^* \in \mathscr{U}$ such that

$$J(u^*) = J^*.$$

3. **Value function and optimality principle.** If u' and u'' are in $\mathscr{U}$ and $0 \leqq s \leqq 1$, let $u = u'_s u''$ in $\mathscr{U}^0$ be given by

$$\begin{aligned} u(t, z) &= u'(t, z), \qquad t \leqq s, \\ &= u''(t, z), \qquad t > s. \end{aligned}$$

Since $\rho(u) = \rho_0^s(u')\rho_s^1(u'')$ we have

$$E(\rho(u)) = E(E\{\rho(u)|\mathscr{F}_s\}) = E(\rho_0^s(u')E\{\rho_s^1(u'')|\mathscr{F}_s\}) = E(\rho_0^s(u')) = 1,$$

hence $u \in \mathscr{U}$. Now suppose that the control u is used. Then the expected future cost at time t, conditioned on the information $\mathscr{Y}_t$ available at time t, is

$$\begin{aligned} \psi^{u'u''}(t) &= E^u\left\{\int_t^1 c^u(s)\,ds \,\middle|\, \mathscr{Y}_t\right\} \\ &= E\left\{\rho(u)\int_t^1 c^u(s)\,ds \,\middle|\, \mathscr{Y}_t\right\} \Big/ E\{\rho(u)|\mathscr{Y}_t\} \qquad \text{a.s.} \end{aligned}$$

The second equality is a well-known property of conditional expectations. Now, for $s \geqq t$, $c^u(s) = c^{u''}(s)$, while $\rho(u) = \rho_0^t(u')\rho_t^1(u'')$ so that

$$E\{\rho(u)|\mathscr{Y}_t\} = E\{E\{\rho_0^t(u')\rho_t^1(u'')|\mathscr{F}_t\}|\mathscr{Y}_t\} = E\{\rho_0^t(u')|\mathscr{Y}_t\}$$

does not depend on u''. Therefore

$$\psi^{u'u''}(t) = f^{u'u''}(t)/E\{\rho_0^t(u')|\mathscr{Y}_t\},$$

$$f^{u'u''}(t) = E\left\{\rho_0^t(u')\rho_t^1(u'')\int_t^1 c^{u''}(s)\,ds \,\middle|\, \mathscr{Y}_t\right\}.$$

If $u' = u'' = u$, we use the notation ψ^u, f^u instead of ψ^{uu} and f^{uu}. Now since c is nonnegative and bounded so is the random variable $f^{u'u''}$. In particular $f^{u'u''} \in L^1 = L^1(C, \mathscr{F}, P)$.

Let $\{f^i\}_{i \in I} \subset L^1$. We say that $f \in L^1$ is the *greatest lower bound* (g.l.b.) of the family $\{f^i\}_{i \in I}$ if (i) $f \leqq f^i$ a.s., $i \in I$, and (ii) if $g \in L^1$ satisfies $g \leqq f^i$ a.s., $i \in I$, then $g \leqq f$ a.s. We write

$$f = \bigwedge_{i \in I} f^i.$$

Evidently if a g.l.b. exists it is unique a.s. It is known that if all the $f^i \geqq 0$, then the g.l.b. exists [**12**, p. 302].

For $u \in \mathscr{U}$, $0 \leqq t \leqq 1$, let

$$V^u(t) = \bigwedge_{u' \in \mathscr{U}} f^{uu'}(t),$$

$$W^u(t) = \bigwedge_{u' \in \mathscr{U}} \psi^{uu'}(t) = V^u(t)/E\{\rho_0^t(u)|\mathscr{Y}_t\}.$$

W^u is called the value function of the control u. Since $f^{uu'}(t)$ is $\mathscr{Y}_t$-measurable it follows that both $V^u(t)$ and $W^u(t)$ are $\mathscr{Y}_t$-measurable. Also $W^u(0) = J^*$ for all u. However, for $t > 0$, $W^u(t)$ depends on u. This situation is quite different from the case of complete observations; i.e., $\mathscr{Y}_t = \mathscr{F}_t$, where W^u does not depend on u. The next lemma shows that $W^u(t)$ is the infimum of the expected future cost conditioned on the observation $\mathscr{Y}_t$.

LEMMA 1. *Let $u \in \mathscr{U}$, $0 \leqq t \leqq 1$. Then for any $\varepsilon > 0$, there exists $u' \in \mathscr{U}$ such that*

$$\psi^{uu'}(t) < W^u(t) + \varepsilon \quad a.s.$$

The proof uses an induction argument based on the following fact. Let u', u'' be in $\mathscr{U}$. Since $\psi^{uu'}(t)$ and $\psi^{uu''}(t)$ are $\mathscr{Y}_t$-measurable it follows that the set

$$M = \{z|\psi^{uu'}(t) < \psi^{uu''}(t)\} \in \mathscr{Y}_t.$$

In turn this implies that the control v, given by

$$\begin{aligned} v(s, z) &= u(s, z), & s &\leqq t, \\ &= u''(s, z), & s &> t, z \notin M, \\ &= u'(s, z), & s &> t, z \in M, \end{aligned}$$

is in $\mathscr{U}$. It is immediate that

$$\psi^{uu'}(t) = \bigwedge \{\psi^{uu'}(t), \psi^{uu''}(t)\} \quad \text{a.s.}$$

To prove the lemma this argument needs to be extended further since we are seeking the g.l.b. of an uncountable set $\{\psi^{uu'}(t)|u' \in \mathscr{U}\}$.

As an immediate consequence of the lemma we get the important result opposite.

THEOREM 2 (PRINCIPLE OF OPTIMALITY). *Let $u \in \mathcal{U}$, $0 \leqq t \leqq 1$, $0 \leqq h$. Then*

$$W^u(t) \leqq E^u\left\{ \int_t^{t+h} c^u(s)\, ds \,\middle|\, \mathcal{Y}_t \right\} + E^u\{W^u(t+h)|\mathcal{Y}_t\} \quad a.s.$$

Furthermore u is optimal if and only if equality holds for all t, h.

COROLLARY 1. *u is optimal if and only if $W^u(t) = \psi^u(t)$.*

Recall that a real-valued process $\{\xi(t)\}$ on $(C, \mathcal{F}, P^u)$, adapted to the family $\{\mathcal{Y}_t\}$ is a *martingale* (*supermartingale*) if

$$E^u\{\xi(t+h)|\mathcal{Y}_t\} = \xi(t) \quad \text{a.s.} \qquad (E^u\{\xi(t+h)|\mathcal{Y}_t\} \leqq \xi(t) \quad \text{a.s.}).$$

We will say that $\{\xi(t)\}$ is a martingale (supermartingale) on $(C, \mathcal{Y}_t, P^u)$.

Suppose that $u \in \mathcal{U}$ is optimal. Then, from Theorem 2,

$$E^u\{W^u(t+h)|\mathcal{Y}_t\} - W^u(t) = -E^u\left\{ \int_t^{t+h} c^u(s)\, ds \,\middle|\, \mathcal{Y}_t \right\} \leqq 0 \quad \text{a.s.},$$

since $c \geqq 0$. Hence $\{W^u(t)\}$ is a supermartingale on $(C, \mathcal{Y}_t, P^u)$.

4. **Representation of supermartingales and optimality conditions.** Let $u \in \mathcal{U}$. We partition the differential equation (1) as

$$dz(t) = \begin{bmatrix} dx(t) \\ dy(t) \end{bmatrix} = \begin{bmatrix} g_1^u(t) \\ g_2^u(t) \end{bmatrix} dt + \begin{bmatrix} \sigma_1(t)dB(t) \\ \sigma_2(t)dB(t) \end{bmatrix}.$$

First of all, we note that the symmetric matrix $\{\sigma_2(t)\sigma_2'(t)\}$ is adapted to $\{\mathcal{Y}_t\}$. To see this, for $\theta \in R^m$, let $\xi(t) = \theta' y(t)$. Then by Itô's differentiation rule [**13**, p. 147] applied to the function ξ^2 we get

$$2\int_0^t \theta' \sigma_2(s)\sigma_2'(s)\theta\, ds = \xi^2(t) - \xi^2(0),$$

and since the right-hand side is $\{\mathcal{Y}_t\}$ adapted, so is the left. Hence there exists a unitary matrix $\{Q(t)\}$ and a positive diagonal matrix $\{L(t)\}$, both adapted to $\{\mathcal{Y}_t\}$, such that $\sigma_2(t)\sigma_2'(t) = Q(t)L(t)Q'(t)$. Set

$$T(t) = (L(t))^{-1/2}Q'(t),$$

$$\hat{g}_2^u(t) = E^u\{g_2^u(t)|\mathcal{Y}_t\}, \qquad \tilde{g}_2^u(t) = g_2^u(t) - \hat{g}_2^u(t).$$

The process $\{v(t)\}$ defined by

$$v(t) = \int_0^t T(s)(dy(s) - \hat{g}_2^u(s)ds)$$

is called the *innovations process* of the observations. The utility of this concept is characterized in the following fact.

THEOREM 3. (i) $\{v(t)\}$ *is a Brownian motion on* $(C, \mathscr{Y}_t, P^u)$.

(ii) *Suppose* $\{\xi(t)\}$ *is a martingale on* $(C, \mathscr{Y}_t, P^u)$. *Then there exists an m-dimensional process* $\{\phi(t)\}$ *adapted to* $\{\mathscr{Y}_t\}$ *with* $\int_0^1 |\phi(s)|^2\, ds < \infty$ *a.s. such that*

$$\xi(t) = \xi(0) + \int_0^T \phi(s)\, dv(s) = \xi(0) + \int_0^T \phi(s)T(s)\,(dy(s) - \hat{g}_2^u(s)\, ds) \quad a.s.$$

The proof of the first part follows from the fact, which is verified by direct calculation, that

$$E^u\{\exp(\theta'(\xi(t) - \xi(s)))|\mathscr{Y}_t\} = \exp(-\tfrac{1}{2}|\theta|^2(t-s)) \quad \text{a.s.}$$

for all $\theta \in R^m$. To prove the second assertion, we start by transforming the measure on $\mathscr{Y} = \mathscr{Y}_1$ from P^u to $\tilde{P}^u$, using Theorem 1, such that $\tilde{y}(t) = \int_0^t T(s)\, dy(s)$ is a Brownian motion on $(C, \mathscr{Y}_t, \tilde{P}^u)$. The measure $\tilde{P}^u$ is given by $\tilde{P}^u(dz) = \pi_0^1 P^u(dz)$, where

$$\text{(4)} \qquad \pi_0^t = \exp\left(\int_0^t - T(s)\hat{g}_2^u(s)\, dv(s) - \frac{1}{2}\int_0^t |T(s)\hat{g}_2^u(s)|^2\, ds\right).$$

Verification of the definition is then used to show that $\{\tilde{\xi}(t) = \xi(t)(\pi_0^t)^{-1}\}$ is a martingale on $(C, \mathscr{Y}_t, \tilde{P}^u)$. A known result [**6**], [**14**], [**15**] yields the representation

$$\text{(5)} \qquad \tilde{\xi}(t) = \int_0^t \tilde{\phi}(s)\, d\tilde{y}(s).$$

From (4) and (5), an application of Itô's differential rule to the product $\xi(t) = \tilde{\xi}(t)\pi_0^t$ gives finally the desired result with

$$\phi(s) = \pi_0^t(\tilde{\phi}(s) - \tilde{\xi}(s)T(s)\hat{g}_2^u(s)).$$

Now suppose that $u \in \mathscr{U}$ is such that $\{W^u(t)\}$ is a supermartingale on $(C, \mathscr{Y}_t, P^u)$. By a well-known result of Meyer [**5**, p. 126], there exists a zero-mean martingale $\{\xi^u(t)\}$ on $(C, \mathscr{Y}_t, P^u)$ and a natural, integrable,[5] process $\{A^u(t)\}$, with increasing sample paths, adapted to $\{\mathscr{Y}_t\}$ such that

$$\text{(6)} \qquad \begin{aligned} W^u(t) &= W^u(0) + \xi^u(t) - A^u(t) \\ &= J^*(0) + \xi^u(t) - A^u(t). \end{aligned}$$

Furthermore $A^u(t)$ is the limit as $h \to 0$ of $A_h^u(t)$ in the weak topology of $L^1(C, \mathscr{Y}_t, P^u)$, where

[5] The technical terms "natural" and "integrable" are defined in [**5**].

(7) $$A_h^u(t) = \int_0^t (W^u(s) - E^u\{W^u(s+h)|\mathscr{Y}_t\})/(h)\, ds.$$

Using (7) we can show in fact that $\{A^u(t)\}$ has absolutely continuous sample paths, so that there exists a nonnegative process $\{\eta^u(t)\}$ such that

(8) $$A^u(t) = \int_0^t \eta^u(s)\, ds.$$

In the decomposition (6) when we use the representation for ξ^u from Theorem 3 and the representation (8) for A^u we obtain

$$\begin{aligned} W^u(t) &= J^* + \int_0^t (\wedge W^u)(s)\, ds + \int_0^t (\nabla W^u)(s)\, dy(s), \\ &= J^* + \int_0^t \left[(\wedge W^u)(s) + (\nabla W^u)(s)\hat{g}_2^u(s) \right] ds + \int_0^t (\nabla W^u)(s) T^{-1}(s)\, d\nu(s) \end{aligned}$$

where the $\{\mathscr{Y}_t\}$-adapted processes $\{(\wedge W^u)(t)\}$, $\{(\nabla W^u)(t)\}$ are given by

$$\begin{aligned} (\wedge W^u)(t) &= -\phi(t)T(t)\hat{g}_2^u(t) - \eta^u(t), \\ (\nabla W^u)(t) &= \phi(t)T(t). \end{aligned}$$

We can now state a main result. Let $\mathscr{U}^s = \{u \in \mathscr{U} | W^u$ is a supermartingale$\}$.

Theorem 4. *$u^* \in \mathscr{U}$ is optimal if and only if*

(i) *there exists a constant V^*,*

(ii) *for each $u \in \mathscr{U}^s$ there exist processes $\{V^u(t)\}$, $\{(\wedge V^u)(t)\}$, and $\{(\nabla V^u)(t)\}$ adapted to $\{\mathscr{Y}_t\}$ such that*

$$\int_0^1 |(\nabla V^u)(t)|^2\, dt < \infty \quad \text{and} \quad E\int_0^1 (\nabla V^u)(t)\, dy(t) = 0 \quad \text{a.s.},$$

$$V^u(t) = V^* + \int_0^t (\wedge V^u)(s)\, ds + \int_0^t (\nabla V^u)(s)\, dy(s), \qquad V^u(1) = 0,$$

and

(iii)
$$\begin{aligned} &(\wedge V^u)^*(t) + (\nabla V^u)(t)\hat{g}_2^{u^*}(t) + c^{u^*}(t) = 0 \quad \text{a.s. } (dt \times dP), \\ &(\wedge V^u)(t) + (\nabla V^{u^*})(t)\hat{g}_2^u(t) + c^u(t) \geqq 0 \quad \text{a.s. } (dt \times dP),\ u \in \mathscr{U}^s. \end{aligned}$$

Furthermore if (i)–(iii) *hold, then $V^* = J^*$ and $V^{u^*} = W^{u^*}$ a.s.*

Proof. The necessity follows immediately from Theorem 2, Corollary

1, and the representation for $\{W^u(t)\}$, $u \in \mathscr{U}^s$, derived above. Conversely, (ii) implies

$$V^u(t) = V^* + \int_0^t [(\bigwedge V^u)(s) + (\nabla V^u)(s)g^u(s)]\,ds + \int_0^t (\nabla V^u)(s)\,d\nu(s),$$

where $\{\nu(t)\}$ is a Brownian motion on $(C, \mathscr{Y}_t, P^u)$. Taking expectations with respect to the measure P^u, and combining with (iii) we get

$$V^* = -E^u \int_0^1 [(\bigwedge V^u)(s) + (\nabla V^u)(s)\hat{g}_2^u(s)]\,ds - E^u \int_0^1 c^u(s)\,ds = J(u),$$

$$V^* = J(u^*).$$

The theorem is proved.

5. **Special cases.** Suppose that the entire vector $z(t)$ is observed so that $\mathscr{Y}_t = \mathscr{F}_t$. Then turning to §3 we see that

$$\psi^{u'u''}(t) = E\left\{\rho_0^t(u')\rho_t^1(u'') \int_t^1 c^{u''}(s)\,ds \middle| \mathscr{F}_t\right\} \Big/ E\{\rho_0^t(u')|\mathscr{F}_t\},$$

$$= \left\{\rho_t^1(u'') \int_t^1 c^{u''}(s)\,ds \middle| \mathscr{F}_t\right\}$$

does not depend on u'. Hence W^u does not depend on u either, and Theorem 4 simplifies to the following.

THEOREM 5. *Suppose $\mathscr{Y}_t = \mathscr{F}_t$. Then $u^* \in \mathscr{U}$ is optimal if and only if*

(i) *there exists a constant V^*,*

(ii) *there exist processes $\{V(t)\}$, $\{(\bigwedge V)(t)\}$ and $\{(\nabla V)(t)\}$ adapted to $\{\mathscr{F}_t\}$ such that*

$$\int_0^1 |(\nabla V)(t)|^2\,dt < \infty, \qquad E\int_0^1 (\nabla V)(t)\,dz(t) = 0 \quad a.s.,$$

$$V(t) = V^* + \int_0^t (\bigwedge V)(s)\,ds + \int_0^t (\nabla V)(s)\,dz(s), \qquad V(1) = 0,$$

and

(iii)
$$(\bigwedge V)(t) + (\nabla V)(t)g^{u^*}(t) + c^{u^*}(t) = 0 \qquad a.s.\ (dt \times dP),$$
$$(\bigwedge V)(t) + (\nabla V)(t)g(t, z, u) + c(t, z, u) \geqq 0, \qquad u \in \mathscr{U},\ a.s.\ (dt \times dP).$$

Furthermore if (i)–(iii) *hold, then $V^* = J^*$ and $V^* = W^*$ a.s.*

REMARK. The optimal control u^* obtained in Theorem 5 is optimal in the "feedback" sense; i.e., for $u \in \mathscr{U}$, $\psi^{uu^*}(t) = W^u(t)$ so that if u^* is used after time t it still minimizes the cost over $[t, 1]$. The same property does not hold in general for the case of partial observations covered by Theorem 4.

We specialize further to the case of diffusion processes. Let $\mathscr{B}_t \subset \mathscr{F}_t$ be the sub-σ-algebra generated by the functions $z(t)$, $z \in C$. We assume that σ, g, c, in addition to the properties stated earlier, satisfy

$$\sigma(t, \cdot), g(t, \cdot, u) \text{ and } c(t, \cdot, u) \text{ are } \mathscr{B}_t\text{-measurable.}$$

Hence σ, g, c can be represented as functions of the form $\sigma(t, z(t))$, $g(t, z(t), u(t))$, $c(t, z(t), u(t))$. We call the resulting problem a Markov problem.

Suppose once again that $\mathscr{Y}_t = \mathscr{F}_t$. It is intuitively clear, however, that, for the Markov problem, a control can safely ignore all the past of the observations and base the control entirely on the present observation $z(t)$. We shall see that this is indeed the case. Let

$$\mathscr{U}^M = \{u \in \mathscr{U} | u(t, \cdot) \text{ is } \mathscr{B}_t\text{-measurable}\}.$$

The first fact of note is that, for $u \in \mathscr{U}^M$, $\{z(t)\}$ is a Markov process on $(C, \mathscr{F}_t, P^u)$, indeed a strong Markov process. This can be shown by checking directly that

$$E^u\{f(z(T+t)) | \mathscr{F}_T\} = E^u\{f(z(T+t)) | \mathscr{B}(z(T))\}$$

for each bounded Borel function f, stopping time T, and $t \geqq 0$. Here $\mathscr{B}(z(T))$ is the sub-σ-algebra generated by the random variable $z(T)$.

Using the Markov property we can show that, for u, u' in $\mathscr{U}^M$,

$$\psi^{M,uu'}(t) = E^u\{\psi^{uu'}(t) | \mathscr{B}_t\}$$

does not depend on u, and consequently

$$W^{M,u}(t) = \bigwedge_{u' \in \mathscr{U}^M} \psi^{M,uu'}(t)$$

does not depend on u either. Hence there is a measurable function $W^M : [0, 1] \times R^n \to R$ such that

$$W^{M,u}(t)(z) = W^M(t, z(t)) \quad \text{a.s.}$$

We can easily derive the principle of optimality:

$$(9) \quad W^M(t, z(t)) \leqq E^u\left\{\int_t^{t+h} c^u(s)\, ds \,\middle|\, \mathscr{B}_t\right\} + E^u\{W^M(t+h, z(t+h)) | \mathscr{B}_t\} \quad \text{a.s.}$$

for each $u \in \mathscr{U}^M$, $t \geqq 0$, $h \geqq 0$. Furthermore equality holds for $u^* \in \mathscr{U}^M$ if and only if u^* is optimal in $\mathscr{U}^M$; i.e., $J(u^*) = W^M(0, z_0)$, $= J^{*M}$, say.

The next step is to obtain a representation for the supermartingale $\{W^M(t, z(t))\}$. We need a more delicate argument than the one presented in §4 since we wish to show that the processes $\{(\wedge W^M)(t)\}$, $\{(\nabla W^M)(t)\}$ obtained there are adapted to $\{\mathscr{B}_t\}$. We do not discuss the proof, and merely state the final result.

There exist measurable functions $\wedge W^M:[0, 1] \times R^n \to R$, $\nabla W^M:[0, 1] \times R^n \to R^n$ such that

$$\begin{aligned} W^M(t, z(t)) = J^{*M} &+ \int_0^t (\wedge W^M)(s, z(s))\, ds \\ &+ \int_0^t (\nabla W^M)(s, z(s))\, dz(s) \quad \text{a.s.} \end{aligned} \tag{10}$$

Armed with the principle of optimality (9) and the representation (10) the next result is easily obtained. The proof is the same as in Theorem 4.

THEOREM 6. *$u^* \in \mathscr{U}^M$ is optimal if and only if*
(i) *there exists a constant V^*,*
(ii) *there exist measurable functions $V:[0, 1] \times R^n \to R$, $\wedge V:[0, 1] \times R^n \to R$, $\nabla V:[0, 1] \times R^n \to R^n$ such that*

$$\int_0^t |(\nabla V)(t, z(t))|^2\, dt < \infty, \qquad E\int_0^1 (\nabla V)(t, z(t))\, dz(t) = 0 \quad a.s.,$$

$$V(t, z(t)) = V^* + \int_0^t (\wedge V)(s, z(s))\, ds + \int_0^t (\nabla V)(s, z(s))\, dz(s) \quad a.s.,$$

$$V(1, z(1)) = 0 \quad a.s.,$$

and

$$\text{(iii)} \quad \begin{aligned} &(\wedge V)(t, z(t)) + (\nabla V)(t, z(t))g(t, z(t), u^*(t, z(t))) \\ &\qquad\qquad + c(t, z(t), u^*(t, z(t))) = 0 \quad a.s. \\ &(\wedge V)(t, z(t)) + (\nabla V)(t, z(t))g(t, z(t), u) \\ &\qquad\qquad + c(t, z(t), u) \geqq 0, \qquad u \in \mathscr{U}^M, \quad a.s. \end{aligned}$$

Furthermore if (i)–(iii) *hold then $V^* = J^{M*}$, $V = W^M$ a.s. Moreover, $J^{M*} = J^*$, $W^M(t, z(t)) = W(t)$ a.s.*

It is important to note that if the function $V(t, x)$, in Theorem 6, has continuous first and continuous first and second partial derivatives respectively in t and x then, by Itô's differential rule, it follows that the processes $\wedge V$ and ∇V are given by

$$\nabla V(t, x) = \frac{\partial V}{\partial x}(t, x),$$

$$\wedge V(t, x) = \frac{\partial V}{\partial t}(t, x) + \frac{1}{2}\sum_{i,j} \frac{\partial^2 V}{\partial x_i \partial x_j}(t, x)(\sigma\sigma')_{ij}(t, x).$$

6. **Concluding remarks.** The recent developments in the theory of martingales has brought about great strides in the development of the theory of stochastic optimal control. We hope this paper has given some flavor of the kinds of results which can be obtained. Further progress should occur along two directions. On the one hand the nonsingularity of the matrix σ must be relaxed to include practical situations (see [**8**]). On the other hand different system models, i.e., using different "noise" processes, must be considered (see e.g. [**16**]).

REFERENCES

1. M. H. A. Davis and P. P. Varaiya, *Dynamic programming conditions for partially observable stochastic systems*, SIAM J. Control (to appear).

2. I. V. Girsanov, *On transforming a class of stochastic processes by absolutely continuous substitution of measures*, Teor. Verojatnost. i Primenen. **5** (1960), 314–330 = Theor. Probability Appl. **5** (1960), 285–301. MR **24** #A2986.

3. R. Rishel, *Necessary and sufficient dynamic programming conditions for continuous time stochastic optimal control*, SIAM J. Control **8** (1970), 559–571. MR **42** #9036.

4. E. B. Dynkin, *Controlled stochastic processes*, Teor. Verojatnost. i Primenen. **10** (1965), 3–18 = Theor. Probability Appl. **10** (1965), 1–14. MR **30** #3794.

5. P.-A. Meyer, *Probability and potentials*, Blaisdell, Waltham, Mass., 1966. MR **34** #5119.

6. H. Kunita and S. Watanabe, *On square integrable martingales*, Nagoya Math. J. **30** (1967), 209–245. MR **36** #945.

7. W. H. Fleming, *Optimal continuous-parameter stochastic control*, SIAM Rev. **11** (1969), 470–509. MR **41** #9633.

8. R. W. Rishel, *Weak solutions of a partial differential equation of dynamic programming*, SIAM J. Control **9** (1971), 519–528.

9. W. H. Fleming and M. Nisio, *On the existence of optimal stochastic controls*, J. Math. Mech. **15** (1966), 777–794. MR **33** #7170.

10. D. W. Stroock and S. R. S. Varadhan, *Diffusion processes with continuous coefficients*. II, Comm. Pure Appl. Math. **22** (1969), 479–530. MR **40** #8130.

11. T. Duncan and P. Varaiya, *On the solutions of a stochastic control system*, SIAM J. Control **9** (1971), 354–371.

12. N. Dunford and J. T. Schwartz, *Linear operators*. I: *General theory*, Pure and Appl. Math., vol. 7, Interscience, New York, 1958. MR **22** #8302.

13. E. Wong, *Stochastic processes in information and dynamical systems*, McGraw-Hill, New York, 1971.

14. J. M. C. Clark, *The representation of functionals of Brownian motion by stochastic integrals*, Ann. Math. Statist. **41** (1970), 1282–1295. MR **42** #5336.

15. M. Fujisaki, G. Kallianpur and H. Kunita, *Stochastic differential equations for the nonlinear filtering problem*, Osaka J. Math. **9** (1972), 19–40.

16. S. M. Ross, *Applied probability models with optimization applications*, Holden-Day, San Francisco, Calif., 1970. MR **41** #9383.
17. V. E. Benes, *The existence of optimal stochastic control laws*, SIAM J. Control **9** (1971), 446–472.

UNIVERSITY OF CALIFORNIA, BERKELEY

Building and Evaluating Nonlinear Filters[1]

R. S. Bucy

A problem for which a rather complete theory has been developed in the past ten years or so is that of nonlinear filtering. The problem consists of observing a portion of the sample function of a vector valued stochastic process, $z(\cdot, \omega)$, and on the basis of the observations to construct the conditional distribution of a nonlinearly related process, $\mathbf{x}(t, \omega)$, the signal process, given the observations. Interest in this problem developed in the early 1950's, as it was a natural generalization of the filtering problem posed and solved by Wiener in [**5**]. In 1961, the author and Kalman, in [**3**], proposed and solved a more general linear filtering problem than that of Wiener, which had, as its starting point, a *linear* stochastic differential equation governing the signal and observation processes and it was shown that the conditional distribution was determined by the solution of an ordinary differential equation for the error covariance and the solution of a stochastic differential equation for the estimate; see [**1**] and [**2**]. This theory has been widely applied in practical guidance system design in the 1960's.

In the early and mid '60's, the full nonlinear problem was solved in [**3**], [**4**] and [**2**]. For this problem, the signal and observation processes were modeled as the solutions of stochastic differential equations. Unfortunately for application, the solution was a Feynman-type path integral, in general not determined by the solution of a finite number of stochastic differential equations. In this paper we will be interested in the

---

AMS (MOS) subject classifications (1970). Primary 93E10, 93-02, 60H10; Secondary 93-04, 94A05.

[1] This research was supported by the United States Air Force, Office of Aerospace Research, under grant AF-AFOSR-71-2141.

approximate solution of a coupled infinite-dimensional set of stochastic differential equations which characterizes the solution of nonlinear filtering problems. In order to produce solutions to nonlinear filtering problems, we use a third-generation digital computer as the synthesis tool.

This talk will review the theory of the continuous and discrete time nonlinear filtering and describe the floating grid and orthogonal function methods of synthesizing nonlinear filters. Finally, some recent results of applying these synthesis techniques to two problems, the passive receiver and the phase demodulator, will be presented, along with a discussion of Monte Carlo evaluation of the optimal nonlinear filter performance.

For details, see [**6**].

REFERENCES

1. R. S. Bucy, *Linear and nonlinear filtering*, Proc. IEEE **58** (1970), 854–864.

2. R. S. Bucy and P. D. Joseph, *Filtering for stochastic processes with applications to guidance*, Interscience Tracts in Pure and Appl. Math., no. 23, Interscience, New York, 1968. MR **42** #2846.

3. R. E. Kalman and R. S. Bucy, *New results in linear filtering and prediction theory*, Trans. ASME Ser. D.J. Basic Engrg. **83** (1961), 95–108. MR **38** #3076.

4. H. J. Kushner, *On the differential equations satisfied by conditional probability densities of Markov processes, with applications*, J. Soc. Indust. Appl. Math. Ser. A Control **2** (1964), 106–119. MR **31** #4642.

5. N. Wiener, *Extrapolation, interpolation and smoothing of stationary time series. With engineering applications*, M.I.T. Press, Cambridge, Mass.; Wiley, New York; Chapman & Hall, London, 1949. MR **11,** 118.

6. R. Bucy, K. D. Senne and C. Hecht, *An engineer's guide to building non-linear filters*, Frank J. Seiler Res. Lab., Air Force Systems Command, Report #SRL-TR-72-0004, May 1972.

UNIVERSITY OF SOUTHERN CALIFORNIA

Wiener's Theory of Nonlinear Noise[1]

H. P. McKean

1. **Introduction.** A (real) process $\mathfrak{x}(t): -\infty < t < \infty$ with $E(\mathfrak{x}) = 0$ and $E(\mathfrak{x}^2) < \infty$ whose statistics are unchanged by the shift $\mathfrak{x} \to \mathfrak{x}(\cdot + t)$ is called "noise". Wiener [**24**] liked to think of such a noise as the output of a "black box" $\mathfrak{f}$, as in Figure 1: You put in a "white noise" $\dot{b}$, and you get $\mathfrak{x}(0)$ out.

$$b \to \boxed{\mathfrak{f}} \to \mathfrak{x}$$

FIGURE 1

The noise $\mathfrak{x}(t)$ is produced by shifting the input by the (incompressible) flow of the white noise $\dot{b} \to \dot{b}(\cdot + t)$. Wiener [**22**] proved that any functional $\mathfrak{f}$ of the white noise with $E(\mathfrak{f}^2) < \infty$ can be expanded into a "power series"

$$\mathfrak{f} = E(\mathfrak{f}) + \sum_{n=1}^{\infty} \int f_n \, d^n b$$

of mutually perpendicular "polynomials"

$$\int f_n \, d^n b = \int_{-\infty}^{\infty} \int_{-\infty}^{t_1} \cdots \int_{-\infty}^{t_n} f_n(t_1, \ldots, t_n) \dot{b}(t_1) \cdots \dot{b}(t_n) \, d^n t$$

of degrees $n = 0, 1, 2, 3$, etc. The "coefficients" f_n are sure functions with

$$\| f_n \|^2 = \int_{-\infty}^{\infty} \int_{-\infty}^{t_1} \cdots \int_{-\infty}^{t_{n-1}} f_n^2(t_1, \ldots, t_n) \, d^n t < \infty,$$

AMS (MOS) subject classifications (1970). Primary 60G10.

[1] This work was supported by the National Science Foundation grant GP-34620.

and

$$E(\mathfrak{f}^2) = E(\mathfrak{f})^2 + \sum_{n=1}^{\infty} \|f_n\|^2.$$

The noise $\mathfrak{x}$ produced by shifting the incoming white noise of Figure 1 can now be expressed as

$$\mathfrak{x}(t) = [E(\mathfrak{x}) = 0] + \sum_{n=1}^{\infty} \int_{-\infty}^{\infty} \int_{-\infty}^{t_1} \cdots \int_{-\infty}^{t_{n-1}} f_n(t_1 - t, \ldots, t_n - t)\, d^n b.$$

The purpose of this paper is to review what is known about such presentations of a fixed noise $\mathfrak{x}$. The name "presentation" is used to signify that you do not much care about the relationship of $\mathfrak{x}$ to $\dot{b}$ *pathwise*; all you want is that $\mathfrak{x}$ and

$$\mathfrak{x}'(t) = \sum_{n=1}^{\infty} \int f_n(\cdot - t)\, d^n b$$

exhibit *the same statistics.*

I think it is fair to say that the subject is not much advanced from the point at which Wiener [**24**] left it; in particular, practical applications have been limited by the lack of a really satisfactory mathematical theory; see, however, Canavan [**2**], Marmarelis and Naka [**12**], and Wiener [**24**] for a sample. The purpose of this review is to indicate some of the obstacles that stand in the way of this goal towards which Wiener so earnestly labored most of his life.

2. **Outline.** There are four main questions as to the presentability of a fixed noise $\mathfrak{x}$.

QUESTION 1 is to ask *can $\mathfrak{x}$ be presented at all?* This is *not* automatic. The theorem of Nisio [**13**] and Wiener [**22**] states that any metrically transitive noise can be approximated (in law) by a presentable noise (see §5 below), but for an exact presentation $\mathfrak{x}$ has to be not just metrically transitive but also *mixing* in the sense that

$$\lim_{T \uparrow \infty} P(A \cap B_T^+) = P(A)P(B),$$

in which B_T^+ denotes the event that the shifted noise $\mathfrak{x}(\cdot + T)$ belongs to B. To see this, let $\mathfrak{e}$ be a functional of $\mathfrak{x}$ (such as the indicator of A or B) with $E(\mathfrak{e}^2) < \infty$. By the presentation of $\mathfrak{x}$, you can expand $\mathfrak{e}$ as a sum

$$\mathfrak{e} = E(\mathfrak{e}) + \sum_{n=1}^{\infty} \int e_n\, d^n b,$$

and if you designate by $\mathfrak{e}_T^+$ the same functional of the shifted path, you will have

$$E(\mathfrak{e}\,\mathfrak{e}_T^+) = E(\mathfrak{e})^2 + \sum_{n=1}^{\infty} (n!)^{-1} \int_{R^n} e_n(t) e_n(t - T)$$

$$= E(\mathfrak{e})^2 + \sum_{n=1}^{\infty} (n!)^{-1} \int_{R^n} \exp(2\pi(-1)^{1/2}(k_1 + \cdots + k_n)T) |\hat{e}_n(k)|^2 \, d^n k,$$

in which $\hat{e}_n$ is the Fourier transform of e_n regarded as a symmetric function on R^n:

$$\hat{e}_n(k) = \int_{R^n} \exp(2\pi(-1)^{1/2} k \cdot t) \, e_n(t) \, d^n t.$$

But then

$$E(\mathfrak{e}\,\mathfrak{e}_T^+) = E(\mathfrak{e})^2 + \int_{-\infty}^{\infty} \exp(2\pi(-1)^{1/2} kT) \Delta(k) \, dk$$

with the summable weight

$$\Delta(k) = \sum_{n=1}^{\infty} (n!)^{-1} |\hat{e}_n(k)|^2,$$

and so

$$\lim_{T \uparrow \infty} E(\mathfrak{e}\,\mathfrak{e}_T^+) = E(\mathfrak{e})^2.$$

The mixing property of $\mathfrak{x}$ is now self-evident. The problem of presentability is part of the classification of incompressible flows; its depth may be judged by the fact that the classification of Bernoulli flows (coin-tossing and the like) has only recently been achieved; see Smorodinsky [**20**] for an exposition and additional information.

QUESTION 2 for a presentable noise is to ask *how much freedom do you have in the actual presentation, i.e., which features of the coefficients* f_n *are essential to the statistics and which are merely accidental*? The simple fact that the statistics of

$$\mathfrak{f} = \sum_{n \leqq d} \int f_n \, d^n b$$

determine the degree d if the latter is finite and $f_n \not\equiv 0$ for $n = d$ already has far-reaching implications; for example, it implies that a Gaussian noise has only transcendental presentations of degree $d \neq 1$; see §8 for the proof and §§9–10 for additional information for degree 2.

QUESTION 3 for a presentable noise is to ask *if you can make the presentation nonanticipating, i.e., can you make* $f_n = 0$ *for* $t_1 \geqq 0$ *so that*

$$\mathfrak{x}(t) = \sum_{n=1}^{\infty} \int_{-\infty}^{t} \int_{-\infty}^{t_1} \cdots \int_{-\infty}^{t_{n-1}} f_n(t_1 - t, \ldots, t_n - t)\, d^n b$$

is independent of the future of $\dot{b}$? The existence of such a presentation implies that the remote past of $\mathfrak{x}$ is trivial: *if* B *is measurable over* $\mathfrak{x}(t)$: $t \leqq T$ *for every* $T < \infty$, *then either* $P(B) = 0$ *or* $P(B) = 1$; in fact, $\mathfrak{x}$ simply inherits this property from the white noise just as it does the mixing property. The conjecture is that *the triviality of the remote past suffices for a nonanticipating presentation.* The evidence to date is favorable but the issue is not yet decided even for chains; see §§4, 6, and 7 for additional information.

QUESTION 4 for nonanticipating noise is to ask *if the presentation can be made nonsingular, i.e., can you make the field of* $\mathfrak{x}(t): t \leqq 0$, *which is always a subfield of the past of the white noise, match the latter exactly*? The importance of this possibility stems from the prediction problem which is to compute the distribution of $\mathfrak{x}(t)$ at some future time ($t > 0$) conditional upon the past $\mathfrak{x}(s): s \leqq 0$.

3. **Gaussian noise.** Questions 1–4 can be answered in a completely satisfactory way for Gaussian noise; the ideas come from Wiener [**23**].

Question 1 is elementary. A Gaussian noise $\mathfrak{x}$ is presentable if and only if it has a (summable) spectral density Δ:

$$E[\mathfrak{x}(0)\mathfrak{x}(t)] = \int_{-\infty}^{\infty} \exp(2\pi(-1)^{1/2}kt)\Delta(k)\, dk.$$

Δ is nonnegative and even, so you can find a function $\hat{f} \in L^2(R^1)$ with $\hat{f}(k)^* = \hat{f}(-k)$ and $|\hat{f}|^2 = \Delta$. The inverse Fourier transform

$$f(t) = \int_{-\infty}^{\infty} \exp(-2\pi(-1)^{1/2}kt)\hat{f}(k)\, dk$$

is real, and

$$\mathfrak{x}'(t) = \int_{-\infty}^{\infty} f(t_1 - t)\dot{b}(t_1)\, dt_1$$

is a presentation of $\mathfrak{x}$, namely, $\mathfrak{x}'$ is also Gaussian with the same spectral density as $\mathfrak{x}$:

$$E[\mathfrak{x}'(0)\mathfrak{x}'(t)] = \int_{-\infty}^{\infty} f(t_1)f(t_1 - t)\, dt_1 = \int_{-\infty}^{\infty} \exp(2\pi(-1)^{1/2}kt)|\hat{f}(k)|^2\, dk.$$

Question 2 is also easily answered. The statistics of $\mathfrak{x}$ are completely specified by $\Delta = |\hat{f}|^2$, so you may change the phase of $\hat{f}$ at pleasure, provided only that you respect the reality condition $\hat{f}(k)^* = \hat{f}(-k)$;

i.e., you may apply to f any orthogonal transformation of $L^2(R^1)$ that commutes with translations.

Question 3 is the same as asking if it is possible to make $f = 0$ on the half-line $t_1 \geqq 0$ without spoiling the "gain" $|\hat{f}|$. The answer is contained in the classical theorem of Paley and Wiener [**15**] which states that this can be done iff

$$\int_{-\infty}^{\infty} (1 + k^2)^{-1}\lg \Delta(k)\, dk = \int_{-\infty}^{\infty} (1 + k^2)^{-1} \lg|\hat{f}(k)|^2\, dk > -\infty.$$

By a fortunate circumstance, this is the same as saying that *the remote past of $\mathfrak{x}$ is trivial.* The proof goes back to Szegö [**21**], Kolmogorov [**9**], and Wiener [**23**]; for a modern proof employing the language of Hardy functions, see Rozanov [**19**], and for an elementary account of the latter, see Dym and McKean [**3**].

Question 4 is equivalent to asking if it is possible to make a further adjustment of f so as to make the one-sided translates $f(\cdot - t)$: $t \leqq 0$ span $L^2(-\infty, 0]$, still keeping $|\hat{f}|^2 = \Delta$; the fact is that this can always be done under the condition

$$\int_{-\infty}^{\infty} (1 + k^2)^{-1}\lg \Delta(k)\, dk > -\infty$$

for the triviality of the remote past. For $f = 0$ on $[0, \infty)$, $\hat{f}(k)$ is analytic on the open lower halfplane and can be expressed there as the product of an "outer" function

$$f_{\text{outer}}(k) = \exp\left[\frac{1}{\pi(-1)^{1/2}} \int_{-\infty}^{\infty} \frac{\lg|\hat{f}(k')|}{k - k'}\, dk'\right]$$

and an "inner" function which is of modulus $=1$ on the axis and $\leqq 1$ below. The inner factor can be chosen at pleasure without spoiling the gain $|\hat{f}|$ on the axis, and it turns out that you achieve the desired spanning of $L^2(-\infty, 0]$ iff you put the inner factor $\equiv 1$. The prediction problem of Kolmogorov [**9**] and Wiener [**23**] is easily solved using this (nonsingular) presentation: For fixed $t > 0$, the distribution of

$$\mathfrak{x}(t) = \int_{-\infty}^{t} f(t_1 - t)\dot{b}(t_1)\, dt_1,$$

conditional upon the past $\mathfrak{x}(s)$: $s \leqq 0$, is Gaussian and, as such, is specified by the conditional moments

$$\mathfrak{m} = E[\mathfrak{x}(t)|\mathfrak{x}(s): s \leqq 0] = \int_{-\infty}^{0} f(t_1 - t)\dot{b}(t_1)\, dt_1$$

and

$$E[|\mathfrak{x}(t) - \mathfrak{m}|^2 | \mathfrak{x}(s): s \leqq 0] = E\left[\left|\int_0^t f(t_1 - t)\dot{b}(t_1)\,dt_1\right|^2\right] = \int_0^t f^2\,dt_1.$$

The evaluations are immediate from the matching of the fields of $\mathfrak{x}(t)$, $s \leqq 0$ and of $\dot{b}(s): s \leqq 0$; see Rozanov [**19**] for more details and additional information. Robinson [**17**] discovered a very beautiful description of the function $f = [\hat{f}_{\text{outer}}]^\vee$ employed above: *In the class of functions* $f = (\hat{f})^\vee$ *with the fixed gain* $\Delta^{1/2} = |\hat{f}|$, *the special choice* $f = [\hat{f}_{\text{outer}}]^\vee$ *makes* $\int_0^t f^2$ *as big as it can possibly be, simultaneously for every* $t \geqq 0$.

4. **Chains.** Questions 1–4 may also be posed for chains $\mathfrak{x}_n, n = \ldots, 0, 1, 2, \ldots$. The subject originates with P. Lévy [**10**]. The role of the white noise $\dot{b}$ is now played by independent "innovations" a_n, $n = \ldots, 0, 1, 2, \ldots,$ with common uniform distribution

$$P(x \leqq a_0 < y) = y - x, \qquad 0 \leqq x < y \leqq 1.$$

The analogue of Q3 is to ask if the chain can be presented as

$$\mathfrak{x}_n = \mathfrak{f}(a_n, a_{n-1}, a_{n-2}, \ldots)$$

with

$$E(\mathfrak{f}^2) = \int_0^1 \int_0^1 \int_0^1 \cdots \mathfrak{f}^2(a_0, a_1, a_2, \ldots)\,da_0\,da_1\,da_2 \ldots < \infty.$$

P. Lévy's mode of attack [**10**, p. 71] seems still to be the only one available. Bring in the conditional distribution

$$p(x) = P(\mathfrak{x}_0 \leqq x | \mathfrak{x}_{-1}, \mathfrak{x}_{-2}, \ldots)$$

and suppose for simplicity that it has no jumps. Then conditional on $\mathfrak{x}_{-1}, \mathfrak{x}_{-2}, \ldots,$ the "innovation" $a_0 = p(\mathfrak{x}_0)$ is uniformly distributed. As such, it is independent of $\mathfrak{x}_{-1}, \mathfrak{x}_{-2}, \ldots,$ and it is easy to see that $\mathfrak{x}_0$ is a function of $a_0, \mathfrak{x}_{-1}, \mathfrak{x}_{-2}, \ldots$:

$$\mathfrak{x}_0 = \mathfrak{f}_0(a_0, \mathfrak{x}_{-1}, \mathfrak{x}_{-2}, \ldots).$$

The same thing can be done with $\mathfrak{x}_{-1}, \mathfrak{x}_{-2}$, etc. by shifting everything back one step at a time: From $\mathfrak{x}_{-1}$ you get a uniformly distributed innovation a_{-1} which is independent of $\mathfrak{x}_{-2}, \mathfrak{x}_{-3}, \ldots$ (and of a_0), $\mathfrak{x}_{-1}$ can be expressed as a function of $a_{-1}, \mathfrak{x}_{-2}, \mathfrak{x}_{-3}, \ldots,$ and this expression can be substituted back into $\mathfrak{x}_0 = \mathfrak{f}_0(a_0, \mathfrak{x}_{-1}, \ldots)$ to obtain a new expression

$$\mathfrak{x}_0 = \mathfrak{f}_1(a_0, a_{-1}, \mathfrak{x}_{-2}, \mathfrak{x}_{-3}, \ldots).$$

Continuing so for n steps you have

$$\mathfrak{x}_0 = \mathfrak{f}_n(a_0, a_{-1}, \ldots, a_{-n}, \mathfrak{x}_{-n-1}, \mathfrak{x}_{-n-2}, \ldots).$$

Now the natural idea is to make $n \uparrow \infty$ so as to get an expression for $\mathfrak{x}_0$ in terms of innovations only

$$\mathfrak{x}_0 = \mathfrak{f}(a_0, a_{-1}, a_{-2}, \ldots);$$

this formula could be shifted to provide a nonanticipating presentation of the whole chain $\mathfrak{x}$. Naturally, for this to be possible, the remote past of $\mathfrak{x}$ will have to be trivial, and it is the statement of Kallianpur and Wiener [**8**] that this suffices. But the proposed presentation is automatically nonsingular, and the sample space of $\mathfrak{x}$ may be "too simple"; see §6 below for such an example in the context of noise and Rosenblatt [**18**] for chain examples. The statement of Kallianpur and Wiener [**8**] is therefore *incorrect*, though the conjecture still stands, both for chains and noise: *If the remote past of* $\mathfrak{x}$ *is trivial, then it can always be presented in a nonanticipating way provided you do not object to building the presentation on a bigger sample space*, i.e., provided you do not insist upon a *nonsingular* presentation.

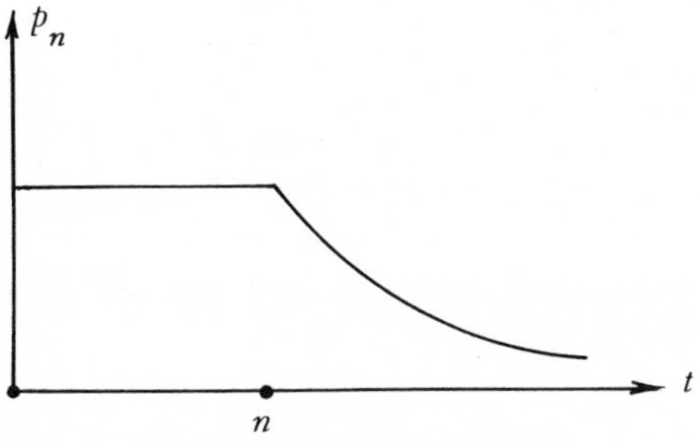

FIGURE 2

5. **Nisio-Wiener's Theorem.** The statement is that *any metrically transitive noise* $\mathfrak{x}$ *can be approximated* (*in law*) *by a presentable noise*. The fact is due to Wiener [**22**]. The proof has been clarified by M. Nisio [**13**], so it is just to call it by both their names.

PROOF. The trick is to produce a series of white noise functionals $\mathfrak{e}_n \geqq 0$ such that

(a) $$P(\mathfrak{e}_n \leqq t) = \int_0^t p_n(s)\, ds$$

with a density function p_n which is constant to the left of $t = n$ and decreasing to the right, as in Figure 2, and

(b) $$(\mathfrak{e}_n)_s^+ = \mathfrak{e}_n - s \quad \text{if } \mathfrak{e}_n \geqq s.$$

Granting that this can be done, the rest is plain sailing. Pick a "typical" sample path $\mathfrak{y}$ of $\mathfrak{x}$ to be specified further below. Then $\mathfrak{f}_n = \mathfrak{y}(-\mathfrak{e}_n)$ satisfies $E(\mathfrak{f}_n^2) < \infty$ because by (a) the density of $\mathfrak{e}_n$ can be expressed as

$$p_n(s) = -\int_n^{\infty} (\text{the uniform density on } s \leqq t)\, dp_n(t)$$

and "typically"

$$\lim_{T\uparrow\infty} \frac{1}{T}\int_{-T}^{0} \mathfrak{y}^2(t)\, dt = E[\mathfrak{x}^2(0)] < \infty.$$

Moreover, the noise

$$\mathfrak{x}_n(t) = \mathfrak{y}[-(\mathfrak{e}_n)_t^+]$$

obtained by shifting $\mathfrak{f}_n = \mathfrak{y}(-\mathfrak{e}_n)$ approximates $\mathfrak{x}$ itself because, by (b), the probability of the cylinder set

$$B = \bigcap_{i=1}^{m} (a_i \leqq \mathfrak{x}_n(t_i) < b_i)$$

is the same as

$$P\left[\bigcap_{i=1}^{m} (a_i \leqq \mathfrak{y}[-(\mathfrak{e}_n)_{t_i}^+] < b_i)\right] = P\left[\bigcap_{i=1}^{m} (a_i \leqq \mathfrak{y}(-\mathfrak{e}_n + t_i) < b_i)\right],$$

up to an error not exceeding $P(\mathfrak{e}_n \leqq \max t_i) = O(n^{-1})$ and, by (a), this is well-approximated for $n \uparrow \infty$ by

$$\lim_{T\uparrow\infty} \frac{1}{T}\int_0^T \text{indicator of} \bigcap_{i=1}^{m} (a_i \leqq \mathfrak{y}(t_i - t) < b_i)\, dt,$$

which is "typically" the same as $P(\mathfrak{x} \in B)$. The proof is finished by producing the functionals $\mathfrak{e}_n$ employed above by the recipie

$$\mathfrak{e}_n = n + \inf Q_n.$$

Q_n stands for the sum of the intervals of length $\geqq n$ of $Q \cap (-n, \infty)$, and Q is the open set on which $b(t) - b(t-1) > 1$. $P(0 \leqq \mathfrak{e}_n < \infty)$ is plain. To verify (a) and (b), you have only to spell out the event $(\mathfrak{e}_n = t)$: If $0 \leqq t < n$, it is the same as saying $t - n \notin Q$ and $(t - n, t) \subset Q$, while if $t \geqq n$, you have to add the proviso $(-n, t - n) \cap Q_n$ *empty*; the latter is automatic from $t - n \notin Q$ if $t < n$. Now for $h \geqq 0$, the event $\mathfrak{e}_n = t + h$ is the same as having $t + h - n \notin Q$, $(t + h - n, t + h) \subset Q$, and $(-n, t + h - n) \cap Q_n$ empty, that is to say

$$(\mathfrak{e}_n = t + h) = ((\mathfrak{e}_n)_h^+ = t) \cap ((-n, h - n] \cap Q_n \text{ empty})$$
$$= ((\mathfrak{e}_n)_h^+ = t) \quad \text{if} \quad t + h < n,$$
$$\subset ((\mathfrak{e}_n)_h^+ = t) \quad \text{in any case.}$$

(a) and (b) are now plain, and the proof is finished.

6. **A singular noise.** By §3, a Gaussian noise $\mathfrak{x}$ with spectral density Δ can always be presented, and the presentation can be made nonanticipating (and also nonsingular) if and only if $\mathfrak{x}$ has a trivial remote past, i.e.,

$$\int_{-\infty}^{\infty} (1 + k^2)^{-1} \lg \Delta(k)\, dk > -\infty;$$

it is to be proved that *a noise may have a nonanticipating presentation, but no nonsingular one.* The example employed is inspired by M. Nisio [**14**].

PROOF. The noise $\mathfrak{x}(t)$ is defined to be $+1$ or -1 according as $b(s) - b(s - 1) > 1$ or < -1 on an interval of length $\geqq 1$ between $t - 2$ and t, and $\mathfrak{x}(t) = 0$ if neither happens. $\mathfrak{x}$ is already presented in a nonanticipating way, but *the field of* $\mathfrak{x}(t), t \leqq 1$, *is so "simple" that only singular presentations exist.* To see this, observe that if $\mathfrak{x}$ is presented in a nonsingular way, then $b(1)$ is measurable over $\mathfrak{x}(t)\colon t \leqq 1$ and is therefore a functional of the latter. But that is plainly impossible as $\mathfrak{x}(t)\colon t \leqq 0$ cannot tell you anything about $b(1)$, and if $\mathfrak{x}(t) = 1$ for $0 \leqq t \leqq 1$ because $b(s) - b(s - 1) > 1$ for $-1 \leqq s \leqq 0$ (which is a nontrivial possibility), no new information is obtained from $\mathfrak{x}$ before time $t = 1$.

The moral is that the triviality of the remote past of $\mathfrak{x}$ is not enough for nonsingular presentation. Be that as it may, *it should guarantee a* (*singular*) *nonanticipating presentation.*

7. **A nonsingular noise.** To have a nice example of a noise that *does* have a nonsingular presentation, pick smooth functions $c_2 > 0$ and c_1 of $-\infty < x < \infty$. If c_1 is of signature opposite to that of x and grows rapidly enough, the diffusion $\mathfrak{x}$ attached to

$$H = (c_2^2/2)\partial^2/\partial x^2 + c_1 \partial/\partial x$$

has a "steady" distribution with density

$$\Delta(x) = \text{a positive multiple of } 2[c_2(x)]^{-2} \exp\left(\int_0^x 2c_1/c_2^2\right),$$

permitting you to define $\mathfrak{x}$ for $-\infty < t < \infty$ with shift-invariant statistics; if also

$$\int_{-\infty}^{\infty} x\Delta\, dx = 0 \quad \text{and} \quad \int_{-\infty}^{\infty} x^2\Delta\, dx < \infty,$$

then $\mathfrak{x}$ is a "noise". Itô and Nisio [**6**] verified that $\mathfrak{x}$ can be presented in a nonsingular way if you employ the white noise $\dot{b}$ defined by $\dot{\mathfrak{x}} = c_2(\mathfrak{x})\dot{b} + c_1(\mathfrak{x})$.

ROUGH PROOF. Pick $T > -\infty$ and $\mathfrak{x}(T)$ and fix them. Then for $t \geqq T$, $\mathfrak{x}$ is found by solving $\dot{\mathfrak{x}} = c_2(\mathfrak{x})\dot{b} + c_1(\mathfrak{x})$ starting from $\mathfrak{x}(T)$ at time $t = T$. This permits you to think of $\mathfrak{x}(t)$ as a nonanticipating functional of the white noise $\dot{b}(t)$: $t \geqq T$ and to expand it into a sum

$$\mathfrak{x}(t) = \mathfrak{x}(T) + \sum_{n=1}^{\infty} \int_T^t \int_T^{t_1} \cdots \int_T^{t_{n-1}} f_n\, d^n b$$

with coefficients f_n depending upon T and $\mathfrak{x}(T)$ but not upon $\dot{b}(T)$: $t \geqq T$. Now you would like to make $T\downarrow -\infty$, and you have the difficulty that $\lim \mathfrak{x}(T)$ and $\lim f_n$ do not exist. This is overcome by taking averages with regard to T and using the fact that

$$\lim_{T\downarrow -\infty} (-1/T) \int_T^0 f[\mathfrak{x}(T')]\, dT' = \int_{-\infty}^{\infty} f(x)\Delta(x)\, dx$$

for nice functions f; the effect is to make $\mathfrak{x}(T)$ disappear in view of

$$\lim_{T\downarrow -\infty} (-1/T) \int_T^0 \mathfrak{x}(T')\, dT' = \int_{-\infty}^{\infty} x\Delta(x)\, dx = 0,$$

and to guarantee that the final coefficients

$$\lim_{T\downarrow -\infty} (-1/T) \int_T^0 f_n\, dT'$$

are sure functions. The presentation so obtained is automatically nonsingular.

The result may be amplified by computing the coefficients of this presentation from the recipie:

$$\begin{aligned} f_n(t_1, \ldots, t_n) &= E[\mathfrak{x}(0)\dot{b}(t_1)\cdots\dot{b}(t_n)] \quad && \text{if } 0 > t_1 > \cdots > t_n, \\ &= 0 && \text{otherwise.} \end{aligned}$$

The formula is

$$f_n = \int_{-\infty}^{\infty} c_2[\exp((t_{n-1} - t_n)H)$$

$$\cdot c_2[\ldots c_2[\exp((t_1 - t_2)H)c_2 \cdot [\exp(-t_1 H)x]']' \ldots]']'\Delta(x)\, dx,$$

in which the primes denote differentiation with regard to x.

PROOF. The chief step is to check that if f is a nice function of $-\infty < x < \infty$, if $t > t_1 > t_2$, and if $F(x) = \exp((t - t_1)H)f(x)$, then

$$E[f[\mathfrak{x}(t)]\dot{b}(t_1)|\dot{b}(s):s \leqq t_2] = E[c_2F'[\mathfrak{x}(t_1)]|\dot{b}(s):s \leqq t_2].$$

To see this you just compute with $t_0 = t_1 + h$ and, for a while, $F = \exp((t - t_0)H)f$:

$$\begin{aligned}
&E[f[\mathfrak{x}(t)]\dot{b}(t_1)|\dot{b}(s):s \leqq t_2] \\
&\quad = \lim_{h\downarrow 0} h^{-1}E[f[\mathfrak{x}(t)] \times b(t_0) - b(t_1)|\dot{b}(s):s \leqq t_2] \\
&\quad = \lim_{h\downarrow 0} h^{-1}E[F[\mathfrak{x}(t_0)] - F[\mathfrak{x}(t_1)] \times b(t_0) - b(t_1)|\dot{b}(s):s \leqq t_2] \\
&\quad = \lim_{h\downarrow 0} h^{-1}E\left[\int_{t_1}^{t_0} F'(\mathfrak{x})\,d\mathfrak{x} + \tfrac{1}{2}\int_{t_1}^{t_0} F''(\mathfrak{x})(d\mathfrak{x})^2 b(t_0) - b(t_1)\middle|\dot{b}(s):s \leqq t_2\right] \\
&\quad = \lim_{h\downarrow 0} h^{-1}E\left[\int_{t_1}^{t_0} c_2F'(\mathfrak{x})\,db \times b(t_0) - b(t_1)\middle|\dot{b}(s):s \leqq t_2\right] \\
&\quad = \lim_{h\downarrow 0} h^{-1}E\left[\int_{t_1}^{t_0} c_2F'(\mathfrak{x})\,ds\middle|\dot{b}(s):s \leqq t_2\right] \\
&\quad = E[c_2F'[\mathfrak{x}(t_1)]|\dot{b}(s):s \leqq t_2]
\end{aligned}$$

with $F = \exp((t - t_1)H)f$ in the final line. The rest of the proof is plain sailing. For example,

$$\begin{aligned}
f_2 &= E[\mathfrak{x}(0)\dot{b}(t_1)\dot{b}(t_2)] \\
&= E[c_2[\exp(-t_1H)x]' \text{ evaluated at } \mathfrak{x}(t_1) \times \dot{b}(t_2)] \\
&= E[c_2[\exp((t_1 - t_2)H)c_2[\exp(-t_1H)x]']' \text{ evaluated at } \mathfrak{x}(t_2)] \\
&= \int_{-\infty}^{\infty} c_2[\exp((t_1 - t_2)H)c_2[\exp(-t_1H)x]']'\Delta\,dx,
\end{aligned}$$

as advertised.

I am offering a modest prize to anybody who can see that $\mathfrak{x}$ is Markovian simply by looking at the expansion $\mathfrak{x} = \sum\int f_n\,d^nb$.

8. **Eidlin and Linnik's bounds.** The next task is to study the statistics of a single functional

$$\mathfrak{f} = \int f_n\,d^nb + \text{terms of lower degree}$$

of degree $n < \infty$ with a view to obtaining information about the degree of freedom that you have in presenting a noise $\mathfrak{x}$; special attention is directed to degree $n = 2$ in §§9 and 10 below. The goal of the present article is to prove that *if* $f_n \neq 0$, *then, for* $x \uparrow \infty$, $-\lg P(|\mathfrak{f}| > x)$ *is bounded above and below by positive multiples of* $x^{2/n}$, *i.e.,*

$$\exp(-k_* x^{2/n}) \leqq P(|\mathfrak{f}| > x) \leqq \exp(-k^* x^{2/n}).$$

Eidlin and Linnik [**4**] obtained the same bounds for polynomial functions of a Gaussian family $\mathfrak{x}_1, \mathfrak{x}_2, \ldots, \mathfrak{x}_m$. The moral is that, *as regards the statistics of* $\mathfrak{f}$, *the degree is not accidental; it is specified by the tail of* $P(|\mathfrak{f}| > x)$.

PROOF FOR $\mathfrak{f} = \int f_n d^n b$ ONLY. The expectation $E(\mathfrak{f}^{2p})$, for $p = 1, 2, 3, \ldots$, is computed by identifying the arguments of the $2p$-fold outer product

$$f_n \otimes \cdots \otimes f_n = \prod_{k \leqq 2p} f_n(t_{k1}, \ldots, t_{kn})$$

by pairs t_{im}, t_{jm} with $1 \leqq i \neq j \leqq 2p$ and $1 \leqq m \leqq n$, then integrating np-fold, and finally summing over all ways of making the identifications. There are $[(2p!)/2^p p!]^n$ such ways, each of which contributes an integral $\leqq \| f_n \|^{2p}$, so $E(\mathfrak{f}^{2p})$ does not exceed a constant multiple of $(2p/e)^{np} \| f \|^{2p}$, and it is a simple exercise to infer that

$$E[\exp(k|\mathfrak{f}|^{2/n})] < \infty$$

for small $k > 0$. The self-evident estimate

$$P(|\mathfrak{f}| > x) \leqq \exp(-kx^{2/n}) E[\exp(k|\mathfrak{f}|^{2/n})]$$

proves the stated upper bound. The proof of the lower bound is adapted from Eidlin and Linnik [**4**]; it assumes familiarity with the expansion of $\mathfrak{f}$ as a sum of products of Hermite polynomials:

$$\mathfrak{f} = \sum_{|p| = n} \text{constant} \times H_{p_1}(\mathfrak{e}_1) H_{p_2}(\mathfrak{e}_2) \times \cdots,$$

in which $|p| = p_1 + p_2 + \cdots$, $H_p(x) = (-1)^p \exp(x^2/2) \times$ (the pth derivative of $\exp(-x^2/2)$), $\mathfrak{e}_n = \int e_n \, db$, and e_n: $n \geqq 1$ is a unit perpendicular basis of $L^2(R^1)$; see, for example, McKean [**11**] for details and additional information. Fix $1 \leqq p < \infty$ and think of $\mathfrak{f}$ as a polynomial $\mathfrak{f}(\mathfrak{e}_1, \ldots, \mathfrak{e}_p)$ with coefficients depending upon $\mathfrak{e}_{p+1}, \mathfrak{e}_{p+2}, \ldots$ only, so that you may write

$$P(|\mathfrak{f}| > x) = E\left[\int (2\pi)^{-p/2} \exp(-r^2/2) r^{p-1} \, dr \, do \right],$$

in which the outer expectation is over $\mathfrak{e}_{p+1}, \mathfrak{e}_{p+2}, \ldots, r \geqq 0$ and $o \in S^{p-1}$ are spherical polar coordinates on R^p; and the inner integral is performed over that portion of $R^p = [0, \infty] \times S^{p-1}$ on which $|\mathfrak{f}(ro)| > x$ for fixed $\mathfrak{e}_{p+1}, \mathfrak{e}_{p+2}, \ldots$. Now pick $c > 0$, $1 \leqq p < \infty$, and $0 < R < \infty$ so as to make $|\mathfrak{f}(ro)| > cr^n$ for $r > R$ on a set Q of positive measure M in the space

$S^{p-1} \times R$ of $(o, \mathfrak{e}_{p+1}, \mathfrak{e}_{p+2}, \ldots)$, as you may do since $\mathfrak{f}$ is of degree precisely n. Inside Q, you have $|\mathfrak{f}(ro)| > x$ for $r > (x/c)^{1/n} > R$, and so

$$P(|\mathfrak{f}| > x) \geqq \int_{(x/c)^{1/n}}^{\infty} \exp(-r^2/2) r^{p-1}\, dr \times (2\pi)^{-p/2} M \geqq \exp(-kx^{2/n})$$

for large x and a small positive constant k. The proof is finished.

A few simple examples will illustrate the use of such bounds; most of the facts can be found in Nisio [**14**], but the present proofs are simpler.

EXAMPLE 1. A bounded nonconstant functional (such as the indicator of an event) is transcendental (not of degree $< \infty$) because $P(|\mathfrak{f}| > x) = 0$ for $x \geqq \|\mathfrak{f}\|_\infty$.

EXAMPLE 2. A functional with a Gaussian tail can only be of degree 1 or transcendental. The second possibility is easily substantiated: For instance, if $\mathfrak{e}(t) = +1$ or -1 according as $b(t)$ is positive or negative, then $\mathfrak{f} = \int_0^1 \mathfrak{e}\, db$ is Gaussian but not of degree 1 because

$$E\left[\mathfrak{f} \int f_1\, db\right] = E\left[\int_0^1 \mathfrak{e} f_1\, dt\right] = \int_0^1 E(\mathfrak{e}) f_1\, dt = 0$$

for any sure function f_1.

EXAMPLE 3. A functional with an exponential distribution $P(\mathfrak{f} > x) = e^{-x}$ must be transcendental or of degree 2; in the second case, a look at the formula of §9 shows that $\mathfrak{f}$ can only be of the form

$$1 + \int e_1 \otimes e_1\, d^2 b + \int e_2 \otimes e_2\, d^2 b$$

with unit perpendicular e_1 and e_2.

EXAMPLE 4. An automorphism of the white noise that maps polynomials into polynomials can only be a "rotation", i.e., it must be of the form

$$b(t) \to \int_{-\infty}^{\infty} (\text{indicator of } 0 \leqq t_1 \leqq t)'\, db$$

in which $e \to e'$ is an orthogonal transformation of $L^2(R^1)$. The reason is that such an automorphism preserves degrees and expectations, so that, for any $e \in L^2(R^1)$, $\mathfrak{e} = \int e\, d^1 b$ is mapped thereby into $\mathfrak{e}' = \int e'\, d^1 b$ with a new $e' \in L^2(R^1)$ in such a way as to preserve $\|e\| = \|e'\|$. The result is now self-evident.

9. **Quadratic statistics.** The next topic is to explain how the statistics of a functional of degree 2 can be computed from its presentation

$$\mathfrak{f} = E(\mathfrak{f}) + \int f_1\, d^1 b + \int f_2\, d^2 b.$$

The constant $E(\mathfrak{f})$ is immaterial, so it is put $=0$ from now on. Think of f_2 as the kernel of a symmetric operator K on $L^2(R^1)$. Because $\|f_2\| < \infty$,

K is compact, with unit perpendicular eigenfunctions e_n and corresponding eigenvalues $\gamma_n \neq 0$ subject to $\sum_{n=1}^{\infty} \gamma_n^2 = \|f_2\|^2 < \infty$. You have

$$f_1 = \sum_{n=1}^{\infty} \beta_n e_n + \text{a piece, } e_\infty\text{, perpendicular to } e_n, \qquad 1 \leqq n < \infty,$$

$$f_2 = \sum_{n=1}^{\infty} \gamma_n e_n \otimes e_n,$$

so

$$\mathfrak{f} = \sum_{n=1}^{\infty} \beta_n \mathfrak{e}_n + \mathfrak{e}_\infty + \sum_{n=1}^{\infty} \gamma_n \tfrac{1}{2}(\mathfrak{e}_n^2 - 1),$$

in which $\mathfrak{e}_n = \int e_n \, d^1 b$ for $n \leqq \infty$ and the formula

$$\int e \otimes e \, d^2 b = \tfrac{1}{2}\left[\left(\int e \, d^1 b\right)^2 - \int e^2 \, dt\right]$$

is used to arrive at the second sum. Because the $\mathfrak{e}_n : n \leqq \infty$ are independent and Gaussian, it is a simple matter to check the formula[2]

$$\begin{aligned} &E(\exp((-1)^{1/2} k \mathfrak{f})) \\ &\quad = E(\exp((-1)^{1/2} k \mathfrak{e}_\infty)) \prod E(\exp((-1)^{1/2} k[\beta_n \mathfrak{e}_n + \tfrac{1}{2}\gamma_n(\mathfrak{e}_n^2 - 1)])) \\ &\quad = \exp(-k^2\alpha^2/2) \prod \frac{\exp(-\tfrac{1}{2}[k^2\beta_n^2/(1 - (-1)^{1/2} k\gamma_n) + (-1)^{1/2} k\gamma_n])}{(1 - (-1)^{1/2} k\gamma_n)^{1/2}} \end{aligned}$$

in which $\alpha = \|e_\infty\|$. Because $\sum \beta_n^2 + \sum \gamma_n^2 < \infty$, the product is convergent except at the points $k = -(-1)^{1/2}/\gamma_n$, and it can be continued along every plane path that does not meet one of these. At the place $k = -(-1)^{1/2}/\gamma_n$, it has a singularity of the form

$$\text{constant} \times (1 - (-1)^{1/2} k\gamma)^{-m/2} \exp(-k^2\beta^2/2(1 - (-1)^{1/2} k\gamma)),$$

in which m is the multiplicity of $\gamma = \gamma_n$ and β^2 is the sum of $\beta_n^2 = (f_1, e_n)^2$ over the m associated eigenfunctions. Therefore, the numbers γ_n are known from the statistics, and you can also read off the multiplicity m and the sum β^2 for each eigenspace separately. The number α is also recovered, and now it is clear that *if you want to change f_1 and f_2 but preserve the statistics of $\mathfrak{f}$, then what you have to respect is* (1) *the spectrum of K, and* (2) *the lengths of the projections of f_1 onto the eigenspaces of K (null space included), keeping track of which length is attached to which eigenvalue. An equivalent statement is that you may rotate f_1 freely within each eigenspace (null space included) and then subject both f_1 and f_2 to a common rotation of $L^2(R^1)$.*

[2] The computation is not at all new; see, for instance, Kac and Siegert [7].

The description of the statistics of cubic polynomials looks as if it would be much more complicated: For example, it is not obvious how the Fourier transform

$$\int_{R^n} \exp((-1)^{1/2} \sum c_{ijk} x_i x_j x_k)(2\pi)^{-n/2} \exp(-|x|^2/2)\, d^n x$$

really depends upon the coefficients c_{ijk} of the cubic form, though it is plain that the classical theory of invariants is involved.

10. **Quadratic noise.** A noise $\mathfrak{x}$ presented by means of a functional $\mathfrak{f} = \int f_2\, d^2 b$ can be more concretely expressed as

$$\mathfrak{x}(t) = \sum_{n=1}^{\infty} \gamma_n \tfrac{1}{2}[\mathfrak{x}_n^2(t) - 1],$$

in which $\mathfrak{x}_n$ is the Gaussian noise produced by shifting $\mathfrak{e}_n = \int e_n\, d^1 b$ and the notation is as in §9; the number of $\mathfrak{x}_n$'s is the "rank" of $\mathfrak{x}$. "Rayleigh noise" is a process of this kind; a simple instance is the envelope $\mathfrak{x}_1^2 + \mathfrak{x}_2^2$ of a Gaussian noise $\mathfrak{x}_1$, e_2 being the Hilbert transform of e_1; see, for example, S. O. Rice [**16**].

Denote once more by K the symmetric operator with kernel f_2 and by $O(t)$ the operation of translation by $t\,[e \to e(\cdot - t)]$. The formula of §9 shows that if you know the statistics of $\mathfrak{x}$, then you also know the spectrum of the compact operator

$$c_1 O(t_1) K O^*(t_1) + \cdots + c_n O(t_n) K O^*(t_n)$$

for every choice of $n \geqq 1$, real $c_1, \ldots, c_n$, and $t_1 < \cdots < t_n$, *and vice versa*, and it is easy to prove (by looking at traces of powers) that the same information is contained in

$$\int \exp(2\pi(-1)^{1/2}[x_1(k_1 - k_2) + x_2(k_2 - k_3) + \cdots + x_n(k_n - k_1)])$$
$$\cdot \hat{K}(k_1, -k_2)\hat{K}(k_2, -k_3) \cdots \hat{K}(k_n, -k_1)\, d^n k$$

as a symmetrical function of $x_1, \ldots, x_n$ for every $n \geqq 2$. The case of rank 1 is already nontrivial: $K = e_1 \otimes e_1$ up to a constant multiplier, and the big integral collapses to

$$\int \exp(2\pi(-1)^{1/2}[x_1(k_1 - k_2) + \cdots + x_n(k_n - k_1)])$$
$$\cdot |\hat{e}_1(k_1)|^2 \ldots |\hat{e}_1(k_n)|^2\, d^n k$$
$$= (|\hat{e}_1|^2)^\vee(x_2 - x_1)(|\hat{e}_1|^2)^\vee(x_3 - x_2) \ldots (|\hat{e}_1|^2)^\vee(x_1 - x_n).$$

F. A. Grunbaum [**5**] has proved that *if you know this product (as a* *symmetric function) for every $n \geqq 2$, then you can find $|\hat{e}_1|$, i.e., for the statistics of $\frac{1}{2}(\mathfrak{x}_1^2 - 1)$ it is precisely the modulus of $\hat{e}_1$ that counts, just as for the statistics of $\mathfrak{x}_1$ itself.* The result is false if n is not permitted to become arbitrarily large; this will appear from the proof.

GRUNBAUM'S PROOF. Denote by ψ the function

$$(|\hat{e}_1|^2)^\vee = \int \exp(2\pi(-1)^{1/2}kx)|\hat{e}_1(k)|^2\, dk;$$

it is real and even since $\hat{e}_1(k)^* = \hat{e}_1(-k)$, and it is continuous since $|\hat{e}_1|^2$ is summable. Besides, $\psi(0) = \|\hat{e}_1\|^2 = 1$, so ψ is positive in the vicinity of $x = 0$, permitting you to recover it there from the known product $\psi(x)\psi(-x) = \psi^2(x)$. Define $0 < y < \infty$ to be the supremum of the numbers x such that $\psi(x')$ can be "detected" from the products $\psi(x_2 - x_1) \cdots \psi(x_1 - x_n)$ (as symmetrical functions) for every $0 \leqq x' \leqq x$. The proof is made by checking that *if $y < \infty$, if $0 < x_1, \ldots, x_n$, and if $x_1 + \cdots + x_n < y$, then*

$$\psi(x_1)\cdots\psi(x_n)\psi(y - x_1 - \cdots - x_n) = 0.$$

The point is that this cannot be maintained for $x_1 = \cdots = x_n = (n+1)^{-1}y$ and large n in the face of $\psi(0) = 1$. The proof now proceeds by induction upon $n = 1, 2, 3 \ldots$. Pick $x_0 = 0 < x_1 < y < x_2$, selecting x_2 so as to make $x_2 - x_1 < y$ and so that $\psi(x_2)$ cannot be detected. The product $\psi(x_1 - x_0)\psi(x_2 - x_1)\psi(x_0 - x_2)$ is already symmetrical as it stands, so you know $\psi(x_1)\psi(x_2 - x_1)\psi(x_2)$ and $\psi(x_1)\psi(x_2 - x_1)$ but not $\psi(x_2)$. The only way for that to happen is to have $\psi(x_1)\psi(x_2 - x_1) = 0$, and the desired identity for $n = 1$ follows upon making $x_2 \downarrow y$. Pick $n \geqq 2$, assume the identity holds for $1, 2, 3, \ldots, n-1$, fix $0 < x_1 < \cdots < x_n$ with $x_1 + \cdots + x_n < y$, and fix also x_{n+1} a little to the right of y in such a way that $\psi(x_{n+1})$ cannot be detected. Think of the product $\psi(x_1' - x_0') \cdots \psi(x_0' - x_{n+1}')$ based upon the points

$$\begin{aligned} x_0' &= 0, \\ x_1' &= x_1, \\ x_2' &= x_1 + x_2, \\ &\cdots \\ x_n' &= x_1 + \cdots + x_n, \\ x_{n+1}' &= x_{n+1}, \end{aligned}$$

and make it symmetrical by adding over all permutations of $x_0', \ldots, x_{n+1}'$. The sum breaks up into $A\psi(x_{n+1}) + B$ with known A and B provided only that x_{n+1} is very close to y, and since $\psi(x_{n+1})$ *cannot* be known,

$A = 0$. This is still so for $x_{n+1} = y$, and that choice is made from now on. The goal is to see how the identity

$$C = \psi(x_1)\cdots\psi(x_n)\psi(y - x_1 - \cdots - x_n) = 0$$

sits in $A = 0$. To do this, think of A as a sum over the permutations of $0, \ldots, n+1$ in which 0 and $n+1$ are adjacent: *If* i *and* $i+1$ *are also adjacent for every* $0 \leqq i \leqq n$, *the permutation contributes to the sum a copy of* C, *while if* i *and* $i+1$ *are adjacent only for* $n \geqq i > j$ *and* $0 < j \leqq n$ *is out of line*, as in Figure 3, then the contribution contains the factor $\psi(x_{k+1} + \cdots + x_j)\psi(x_{j+1})\cdots\psi(x_n)$ for some $0 \leqq k < j$, and this can be made to vanish if you multiply it by $\psi(x_1)\cdots\psi(x_k)\psi(y - x_1 - \cdots - x_n)$ seeing as the product so obtained is of the form $\psi(x_1^*)\psi(x_2^*)\cdots\psi(x_m^*) \cdot \psi(y - x_1^* - \cdots - x_m^*)$ with $m < n$. But now you see that after multiplication by C the sum for A can be reduced to a sum over permutations that preserve every adjacency; in short, $0 = AC = 2n \times C^2$, there being $2n$ such permutations. The proof is finished.

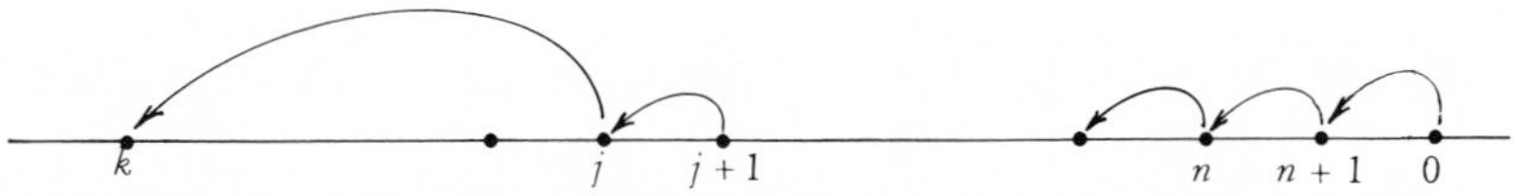

FIGURE 3

PROBABILISTIC PROOF. A second proof is available if $\mathfrak{x}_1$ is sufficiently nice, e.g., if $|\hat{e}_1|$ dies away fast enough. Then $\dot{\mathfrak{x}}_1 \neq 0$ at the (isolated) roots of $\mathfrak{x}_1 = 0$, so that you can recover $\mathfrak{x}_1$ (*up to a single ambiguous signature*) from the modulus $|\mathfrak{x}_1|$ by reflecting every other excursion of the latter, and it follows that $\mathfrak{x}_1(0)\mathfrak{x}_1(t)$ is measurable over $\mathfrak{x}_1^2$. But this means that

$$E[\mathfrak{x}_1(0)\mathfrak{x}_1(t)] = \int \exp(2\pi(-1)^{1/2}kt)|\hat{e}_1(k)|^2\, dk$$

can be found from the statistics of $\mathfrak{x}_1^2$, which is what was wanted.

The first proof suggests a connection with the theorem of Adler and Konheim [**1**]: *If* f *is summable and if you know*

$$\int_{-\infty}^{\infty} f(x)f(x + x_1)\cdots f(x + x_n)\, dx$$

for every $x_1, \ldots, x_n$ *and* $n \geqq 1$, *then you know* f *up to a translation*; it is a nice problem to clarify this. The second proof is also suggestive: The field of $\mathfrak{x}_1$ was seen to be the same as the field of $|\mathfrak{x}_1|$ with a *single* ambiguous signature adjoined, and the conjecture is that *this is still correct even if the paths are badly behaved.*

For rank $\geqq 2$, there is at present only an unproved conjecture: *The statistics of* $\mathfrak{x}$ *are equivalent to the joint statistics of* $\mathfrak{x}_n : n \geqq 1$, *up to individual rotations of the eigenspaces; for instance, if* $e_1, \ldots, e_m$ *span out a single eigenspace, then you cannot tell the statistics of* $\mathfrak{x}_1, \ldots, \mathfrak{x}_m$ *from those of*

$$\mathfrak{x}'_i = \sum_{j \leqq m} o_{ij} \mathfrak{x}_j, \qquad i \leqq m,$$

for any fixed $(o_{ij}) \in O(m)$. The conjecture may also be stated so: *The statistical information is (or should be) the same as knowing* (1) *the eigenfunctions up to individual rotations of the eigenspaces and a common "phase shift", i.e., an orthogonal transformation that commutes with translations, and* (2) *the associated eigenvalues.*

The paper closes with a couple of detection problems closely related to that for $\mathfrak{x} = \frac{1}{2}(\mathfrak{x}_1^2 - 1)$. I owe them to F. A. Grunbaum. The first problem is to detect $\mathfrak{x}_1$ *from* sign $\mathfrak{x}_1$. A moment's reflection will convince you that this cannot be done *pathwise*, but it can be done at the *statistical* level; in fact it is very easy: If $E[\mathfrak{x}_1(0)\mathfrak{x}_1(t)] = \cos\theta$, then

$$P[\mathfrak{x}_1(0) > 0, \mathfrak{x}_1(t) > 0] = \int (2\pi)^{-1} \exp(-|x|^2/2)\, d^2x = (2\pi)^{-1}[\pi - |\theta|],$$

the integral being performed over $x_1 \geqq 0, \cos\theta x_1 + \sin\theta x_2 > 0$. The second problem is to detect $\mathfrak{x}_1$ *from the roots of* $\mathfrak{x}_1 = 0$, which looks even more unlikely as the roots would appear to contain even less information than the signatures. By the root statistics, you will understand

$$\lim_{\varepsilon \downarrow 0} E[\mathfrak{x}_1^*(t_1) \cdots \mathfrak{x}_1^*(t_n)] = (2\pi)^{-n/2} (\det Q)^{1/2},$$

in which x^* is the indicator of $|x| < \varepsilon$ divided by 2ε and

$$Q_{ij} = E[\mathfrak{x}_1(t_i)\mathfrak{x}_1(t_j)].$$

But it is not difficult to see that the information in $(\det Q)^{1/2}$ for $t_1 < \cdots < t_n$ and $n \geqq 1$ is the same as the information in the statistics of $\mathfrak{x}_1^2$, so the statistics of $\mathfrak{x}_1$ itself can be recovered, by Grunbaum's theorem.

REFERENCES

1. R. L. Adler and A. G. Konheim, *A note on translation invariants*, Proc. Amer. Math. Soc. **13** (1962), 425–428. MR **26** #4172.

2. G. H. Canavan, *Some properties of a Lagrangian Wiener-Hermite expansion*, J. Fluid Mech. **41** (1970), 405–412.

3. H. Dym and H. P. McKean, *Fourier series and integrals*, Academic Press, New York, 1972.

4. V. L. Eidlin and Ju. V. Linnik, *A remark on analytic transformation of normal vectors*, Teor. Verojatnost. i Primenen. **13** (1968), 751–754 = Theor. Probability Appl. **13** (1968), 707–710. MR **39** #2202.

5. F. A. Grunbaum, *The square of a Gaussian process*, Z. Wahrscheinlichkeitstheorie und Verw. Gebiete **23** (1972), 121–124.

6. K. Itô and M. Nisio, *On stationary solutions of a stochastic differential equation*, J. Math. Kyôto Univ. **4** (1964), 1–75. MR **31** #1719.

7. M. Kac and A. J. F. Siegert, *On the theory of noise in radio receivers with square law detectors*, J. Appl. Phys. **18** (1947), 383–397. MR **8**, 522.

8. G. Kallianpur and N. Wiener, *Non-linear prediction*, Technical Report No. 1, ONR, Cu-2-56-Nonr-266, (39)-CIRMIP, Proj. NR-047-015, 1956.

9. A. N. Kolmogorov, *Interpolation und Extrapolation von stationären zufälligen Folgen*, Izv. Akad. Nauk SSSR Ser. Mat. **5** (1941), 3–14. (Russian) MR **3**, 4.

10. P. Lévy, *Théorie de l'addition des variables aléatoires*, Gauthier-Villars, Paris, 1937.

11. H. P. McKean, *Geometry of differential space*, Ann. Prob. **1** (1972).

12. P. Marmarelis and K. I. Naka, *White noise analysis of a neuron chain: An application of the Wiener theory*, Science **175** (1972), 1276–1278.

13. M. Nisio, *On polynomial approximation for strictly stationary processes*, J. Math. Soc. Japan **12** (1960), 207–226. MR **22** #6019.

14. ———, *Remarks on the canonical representation of strictly stationary processes*, J. Math. Kyôto Univ. **1** (1961/62), 129–146. MR **26** #7020.

15. R. E. A. C. Paley and N. Wiener, *Fourier transforms in the complex domain*, Amer. Math. Soc. Colloq. Publ., vol. 19, Amer. Math. Soc., Providence, R.I., 1934.

16. S. O. Rice, *Mathematical analysis of random noise*, Bell System Tech. J. **23** (1944), 282–332. MR **6**, 89.

17. E. A. Robinson, *Random wavelets and cybernetic systems*, Griffin's Statistical Monographs and Courses, no. 9, Hafner, New York, 1962. MR **26** #7457.

18. M. Rosenblatt, *Markov processes*, Springer-Verlag, Berlin and New York, 1971.

19. Ju. V. Rozanov, *Stationary random processes*, Fizmatgiz, Moscow, 1963; English transl., Holden-Day, San Francisco, Calif., 1967. MR **28** #2580; MR **35** #4985.

20. M. Smorodinsky, *Ergodic theory, entropy*, Lecture Notes in Math., vol. 214, Springer-Verlag, Berlin and New York, 1971.

21. G. Szegö, *Beiträge zur Theorie der Toeplitzchen Formen*, Math. Z. **6** (1920), 167–202.

22. N. Wiener, *The homogeneous chaos*, Amer. J. Math. **60** (1930), 897–936.

23. ———, *Extrapolation, interpolation, and smoothing of stationary time series. With engineering applications*, M.I.T. Press, Cambridge, Mass.; Wiley, New York; Chapman & Hall, London, 1949. MR **11**, 118.

24. ———, *Nonlinear problems in random theory*, Wiley, New York, 1958. MR **20** #7337.

COURANT INSTITUTE OF MATHEMATICAL SCIENCES, NEW YORK UNIVERSITY